PACKAGING REGULATIONS

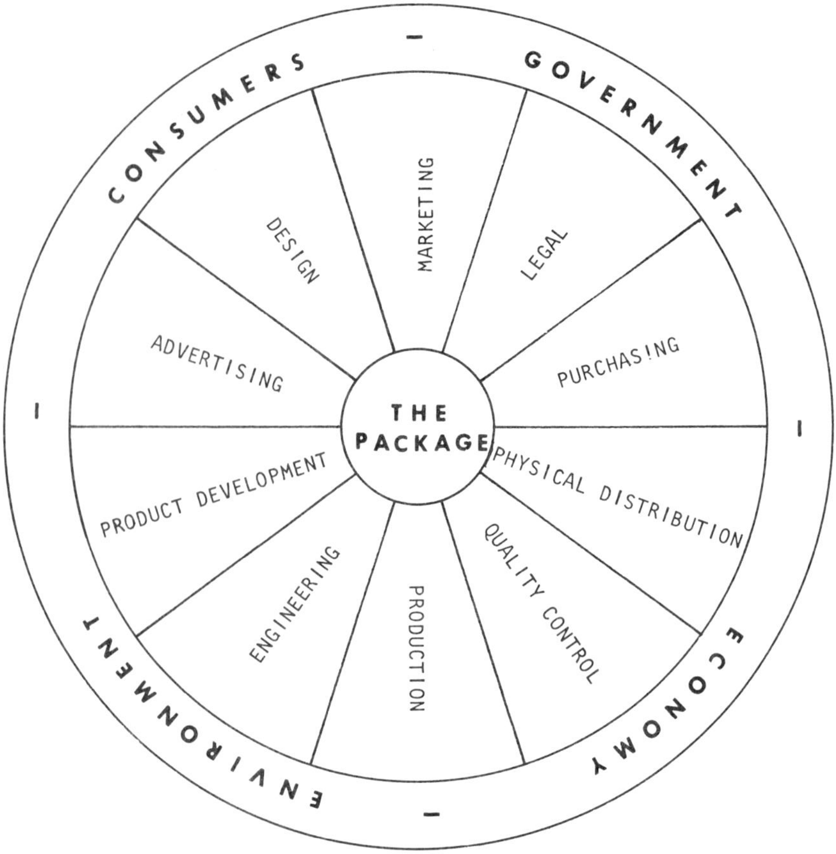

CONSUMERS
GOVERNMENT
ECONOMY
ENVIRONMENT
THE PACKAGE
MARKETING
LEGAL
PURCHASING
PHYSICAL DISTRIBUTION
QUALITY CONTROL
PRODUCTION
ENGINEERING
PRODUCT DEVELOPMENT
ADVERTISING
DESIGN

PACKAGING REGULATIONS

Stanley Sacharow, A.B., A.M.

Professional Member,
The Packaging Institute

AVI PUBLISHING COMPANY, INC.
Westport, Connecticut

Library of Congress Cataloging in Publication Data

Sacharow, Stanley.
Packaging regulations.

Includes index.
1. Packaging — Law and legislation — United States.
2. Packaging — Law and legislation — Canada. I. Title.
KFI665.S3 343'.73'075 78-25786
ISBN 0-87055-274-0

Printed in the United States of America

Preface

During the spring 1977 semester, this writer had the occasion to attend a course dealing with packaging regulations held during the evening session at Rutgers University. Sponsored by the Packaging Science and Engineering (PS&E) curriculum at Rutgers and taught by Mr. Robert J. Kelsey, the course stimulated interest in the vast subject of packaging regulations. The net result is this text. Based on some material presented during the course and an even greater amount obtained by intimate experience in the packaging industry for over 15 years, this text summarizes all regulations dealing with packaging.

The trend in all regulations affecting packaging is towards statutes and the codification of common law. Since a statute dominated system involves considerable governmental responsibility, the text has been divided into the basic governmental agencies' handling of packaging as well as categorized by function. Thus, a chapter on the activities of the FDA precedes a chapter entitled, "Patents, Trademarks and Copyrights." And since all law is divided into religious, common and statute, material is also presented on the religious regulations dealing with the packaging industry.

Statutues change and regulations may be considerably different after a certain period of time. Even though this is true, basic principles hold for years and even centuries. The truths presented in the *Magna Carta* are still valid more than six centuries after their formulation. Perhaps the actual laws may change, but the general principles of governmental regulation are as valid today as they were in the early days of the 20th Century. It is in this vein that this text was written.

Thanks must be given to many. From the various officials who supplied the photos to Bob Kelsey, whose course teased the author's imagination into delving further into a most important topic. From Canada, my friend, Dr. Marvin A. Tung, Associate Professor, Food Science, University of British Columbia and Ms. Maureen R. Garland have written an excellent chapter on Canadian regulations. Their work will prove to be of great value not only to Canadian researchers but to all members of the packaging fraternity. I deeply appreciate their efforts. As in my past books, this work would not be possible without full cooperation of my wife Beverly Lynn and sons, Scott Hunter and Brian Evan. Their devotion and belief in this project made it all worthwhile.

STANLEY SACHAROW
East Brunswick, N.J.

November, 1978

Foreword

As a lawyer, it is with both pleasure and a bit of amicable envy that I write the Foreword to this, Stanley Sacharow's latest fine work in the complex but vitally important packaging field. Lawyers for some time, as a group, have too often been victims of their own jargon and resultant inability to communicate with the public. Their job includes the need to communicate, first and foremost. This Mr. Sacharow has done so well, so admirably. Hence, my "pleasure and amicable envy," for here is a volume easy to read, filled with important and timely knowledge as expounded by an expert and master researcher, knowledge to which the reader has been given ready and quick access by Mr. Sacharow. One can see the author was concerned that what he had to say was not only said, but available without hurdles, to be understood and to be used. Mr. Sacharow has undertaken a demanding but worthwhile project, has seen it through to successful fruition, making the reader and all those concerned with this ambient subject the ultimate beneficiaries.

JAY J. LEVIT, A.B., J.D., LL.M.
Instructor, University of Michigan Law School
Adjunct Associate Professor of Law,
University of Richmond Law School

September 8, 1978

Dedication

To Dr. Harvey Wiley - who started it all!

Contents

"taste cannot be controlled by law"

Thomas Jefferson

"The problem is that the United States has not learned to use science as a regulator, although no other country has either, and the solutions have been based on laws passed in reaction to crises rather than on science."

A Federal Official, *N.Y. Times*
March 15, 1977

1

Packaging Regulations — An Overview

Regulations regarding food and indirectly packaging go back to antiquity. Archaeological discoveries from the Hittite period (3500-1200 B.C.) have proven that even then there was a fairly comprehensive system available which covered foods. Egyptian pictographs have also been unearthed showing various officials measuring and weighing grain. But perhaps the earliest civilizations where food regulations can be directly traced are the Roman and Greek.

In its prime, the Roman Empire ruled almost all of modern-day Europe. Intrinsic to ancient Roman values was not only cleanliness but the value of good commercial practice. For the former, the Romans built baths and concerned themselves with the purity of body as well as food. And for the latter, an elaborate system of weights and measures was devised. Every Roman market had a "Libripen" (Master of the Public Weights) whose function was to observe the weight of the commodities sold and payment received in various transactions.

In the ancient Greek colony of Cyrene (Cyrenaica) in North Africa, archaeologists have uncovered mosaics illustrating King Arcesilas supervising the weighing and storing of silphium (an ancient herb).

In later centuries, Islamic civilization (600-1000 A.D.) contained regulations regarding a weights and measure system which certified various containers as to their proper fill.

But these very early regulations were not directed toward providing a safe and wholesome supply of food for the common man. Rather, they were solely intended to protect each merchant from the other and were used for tax purposes by the ruler. This carried right through into the Middle Ages when Europe was fragmented into small fiefs or city states.

Each charged a tax on its goods sold and each had its own version of food regulations. Bread was stamped in Medieval Europe to certify that the baker had complied with the law. But in spite of these regulations, food adulteration was widespread. Wine and bread were so commonly adulterated during this time that several specific regulations were promulgated covering these products. In virtually all parts of Europe, manufacturers of food products considered dangerous to health were imprisoned or otherwise punished. Yet, severe adulteration of foods continued because of both poor enforcement and widespread ignorance of these laws.

At the end of the Middle Ages and right up into the late 1800's, bread was mixed with alum, beer adulterated by various narcotics, and tea leaves colored with carbonate of copper. The eminent British painter James Hogarth wrote letters to the newspapers of the day rallying against these practices. The situation caused Charles Dickens to write in one of his novels, "If they be like to die, let them do so." And in 1820, Frederick Accum published the first book solely devoted to food adulterations, "A Treatise on Adulterations of Food and Culinary Poisons."

In the United States, the climate relative to proper and meaningful food legislation started to change after the Civil War (1860-1865). Spurred on both by the conservation and human rights movements, it reached its climax during the Spanish-American War (1898-1901). When the battleship "Maine" exploded in Havana, Cuba, U.S. soldiers were subsequently rushed to the battlefront. And feeding these soldiers was as

Courtesy of The Bettmann Archive

FIG. 1.1. SPANISH AMERICAN WAR
Typhoid patient being taken to hospital, Camp Meade, Pa.

important as clothing them. One type of provision supplied to the U.S. Army was canned beef. The beef was purchased by the Army and even before the U.S. soldiers arrived in Cuba, many died in Tampa, Florida, from tainted product. Investigation showed that the beef was pickled in formaldehyde and even canned 30 years before during the Civil War. This so outraged the American public that William Randolph Hearst picked it up and published it in all his newspapers. Public opinion was forming and a group of concerned citizens led by Dr. Harvey Wiley pressed hard for regulations capable of stopping such practices. They did not succeed in capturing general consent until the publication of Upton Sinclair's book, "The Jungle." It was then that the unsanitary conditions prevailing in some meat-packing plants first became known to the public. Theodore Roosevelt recognized the need for improving conditions of these meat plants and pushed for the adoption of food and drug laws. All this culminated in the passage of the first Food and Drug Law in 1906.

FOOD AND DRUG LAW — 1906

This was the first federal Food and Drug Law established in the United States. It was not all encompassing and it did contain many loopholes. There were regulations on label statements relative to composition and identity of food. Various regulatory Standards of Identity were written for some foods and even today some still exist. An example is ketchup. An important aspect of this new regulation was that the government had to prove that there was an unsanitary or poisonous property in the food that was dangerous to health. But a start had been made in the right direction. Administered by the newly formed Bureau of Chemistry under Dr. Wiley, many food processors saw the necessity for this regulation. They supported the act and backed up the Bureau of Chemistry. This was the first time that both government and industry joined to enforce a law.

Between 1907 and 1912, the law only referred to statements on composition or identity. With the passage of the Sherley Amendment in 1912, violators became subject to prosecution. Even though the government had to prove harmfulness, more than 6000 violators were brought to trial during the first ten years of the law. This was in spite of the Supreme Court which refused to consider these cases. They believed them to be in the realm of business and not public health.

In 1930, the McNary-Mapes Amendment was passed. This codified the quantity of contents in packaged foods. Exceptions were usually found in packaging of meat or milk products due to agricultural pressures. An interesting point is that both meats and poultry were not covered in the 1906 Act. (See Table 1.1.) This was in spite of Sinclair's book. Both meat and poultry were covered in the 1906 Meat Inspection Act and its later amendment. Because of this initial omission and also the fact that the Bureau of Chemistry was relatively underfinanced, responsibility for the enforcement of regulations on meats and poultry was given to the USDA.

TABLE 1.1. FOOD DEFINITIONS

A. The 1906 Statutory Definition:
- (a) "All articles used for food, drink, confectionery, or condiment—
- (b) "by man or other animals,—
- (c) "whether simple, mixed, or compound."
- (d) Inclusions and exclusions

B. The 1938 Statutory Definition:
- (a) "Articles used for food or drink—
- (b) "for men or other animals,—
- (c) "chewing gum, and—
- (d) "articles used for components of any such article."

Other regulations soon followed. In 1914, the Federal Trade Commission (FTC) was given the responsibility for the enforcing of textile labeling. Wool was covered in 1941, while fur labeling followed later in 1952. In general, the existing food regulations up until the late 1930's were considered too weak to enforce a good wholesome food supply. As so often in the past, it took a new calamity to force the enactment of improved regulations. This occurred in 1937 with the "Elixir Sulfanilimide" event. When almost 70 people died as a result of taking this new drug, the public became incensed. The net result was the passage of more improved legislation in 1938.

FEDERAL FOOD, DRUG AND COSMETIC ACT — 1938

Often referred to as a "patchwork cathedral to which is added new chapels continually in which to worship new dieties," the 1938 measure was nonetheless truly revolutionary legislation. It has served as a model for other food and drug legislation on a world-wide basis. There are many aspects of this new law that set the pace for future activities. A new agency was set up called the Food and Drug Administration (FDA), and it was made responsible for all administrative decisions. Food additives

Courtesy of U.S. FDA

FIG. 1.2. FDA INSPECTOR AT WORK
In a dockside warehouse on Staten Island, import inspectors collect routine inspection samples of spices and make a preliminary visual inspection before laboratory analysis.

were not controlled except where such additives were known to be poisonous substances. Even this was a significant improvement over the 1906 law since between 1906 and 1938 classification and regulatory laws regarding toxic additives were handled piecemeal as the problems arose. Chemicals prohibited for use in edibles or in materials coming in contact with them were mainly those which had a well established history of toxicity.

The Second World War interfered with the full application of the 1938 Act, but several specific details of the Act are important. Standards of identity and a reasonable standard of quality and of fill for various foods were specified. Traffic in injurious foods was prohibited. More requirements about ingredients on labels were specified. Slack filling was brought under review and the use of deceptive containers was prohibited. Penalties were invoked to control fraud or deliberate intent to violate the act.

The 1938 legislation designed to "provide [food] safe and wholesome to the people, honestly labeled and properly packaged" was the basic law relative to food packaging until the later 1950's. (See Table 1.2.) It was then that external events again caused a new stirring in the public's awareness of packaging safety.

TABLE 1.2. 1938 FEDERAL FOOD, DRUG, AND COSMETIC ACT PROVISIONS

(1) Prohibited foods dangerous to health
(2) Prohibited unsanitary packages
(3) Established FDA with decent budget
(4) Plugged loopholes in 1906 law
(5) Established stiff fines
(6) Plants must stop production until FDA gives permission
(7) Injunctions could be issued against offender by courts
(8) Additives not really covered since not many then available

In 1956, a convention of physicians meeting in Milan, Italy, devoted their seminars to a comprehensive study of cancer and its occurrence. They arrived at several rather startling conclusions one of which was that cancer *may* be caused by the use of various chemical additives in food. There were not too many people who paid attention to the conference's conclusions at the time, except for one U.S. Congressman. Rep. James J. Delaney (*D*—N.Y.) thought that the findings of the Milan convention were quite important and highly significant in the battle for public health. He managed to convince Sen. Estes Kefauver (*D*—Tenn.) and both pushed for the drafting of an amendment to the 1938 Act. Consumerism had just begun to become a vital factor in the public's mind. The net result was new legislation that directly concerned the packaging industry.

1958 FOOD ADDITIVES AMENDMENT (TO THE 1938 FOOD, DRUG AND COSMETIC ACT)

This was the second most revolutionary concept in food legislation for the common man. The amendment deals with food supply in the United States and takes cognizance of all chemical components of this supply, whether these components enter the food by direct addition or by indirect means, for example, by virtue of migration from a food packaging material. (See Table 1.3.)

Courtesy of Rep. Delaney's Office

FIG. 1.3. REP. JAMES J. DELANEY
New York Congressman—9th District.

TABLE 1.3. 1958 FOOD ADDITIVES AMENDMENT PROVISIONS

(1) Additives defined in Amendment
(2) Burden of proof now shifted to industry
(3) Delaney Clause outlawed carcinogens
(4) Established "GRAS" (Substances generally recognized as safe) category
(5) Established GMP (Good manufacturing regulations) category

When the 1958 Food Additive Amendment was passed, the FDA was faced with a great deal of work. It had to find a way to administer the new law as well as bring under control, by scientific evaluation, the many thousands of substances involved in foods and in packaging materials for foods.

The most important provisions of the 1958 Amendment are summarized in Table 1.3; however, one provision is of even greater importance to the packaging industry of the 1970's. This is the clause commonly known as the Delaney Clause — which bars the use of any additive in food that is found "to induce cancer in man or animal." At the time of its inclusion into the 1958 Amendment, the test methods used to detect carcinogenic substances were valid to about 20 parts/million. Advances in testing procedures have now brought these analytical methods to quantitative accuracies of about a few parts/trillion. Testing tolerances have been drastically narrowed since 1958 and researchers are now finding almost everything present everywhere. And yet the Delaney Clause specifically outlaws any additive in any amount found to be carcinogenic. How to cope with this provision is of great concern to the packaging industry at the present time.

TABLE 1.4. EVENTS AND RESULTS

Event	Result
(1) Ten children died of diptheria antitoxin	Biologics Control Act of 1902
(2) Soldiers died as result of tainted canned beef	Food and Drug Law — 1906
(3) "Elixir Sulfanilimide" event — 70 people died	Federal Food, Drug and Cosmetic Act 1938
(4) Physicians meeting in Milan, Italy on cancer	Food additives Amendment to the Federal Food, Drug and Cosmetic Act — 1958
(5) "Thalidomide" incident — birth defects caused by new drug	New Drug Act — 1962 (Public Law 87-871)
(6) Over 2,000 deaths of children due to accidental poisoning	Poison Prevention Packaging Act of 1970
(7) Over 350 reports annually of allergic reactions to cosmetics	Cosmetic Ingredient labeling — voluntary and mandatory provisions
(8) Marketing of useless radiological devices for cancer cure — other useless devices on market, i.e., arthritis cures, etc.	Medical Device Amendment — 1975 (GMP published in late 1977)

While there was all this fervent activity between 1938 and 1958 in the food and drug areas, other legislation was not ignored. (See Table 1.4.)

In 1939, the Federal Seed Act specified the type of labeling information that must appear on seed packages. Other legislative measures in the agricultural area included the Federal Insecticide, Fungicide, and Rodenticide Act of 1947 which controls these type compounds in foods as well as in other commodities. These substances are under FDA jurisdiction by virtue of the above law and are not covered under the provisions of food legislation.

In the 1960's, the Dangerous Cargo Act was passed. This act, coupled with the Hazardous Substances Labeling Act, broadened the Caustic Provision Act of 1921 to include all industrial consumer products. The Child Protection Act soon followed.

In 1976, the Hazardous Substances Labeling Act, which had been administered by the various shipping groups (not the FDA) such as the Coast Guard came under the jurisdiction of the Department of Transportation (DOT). At present, the DOT is attempting to codify the various materials involved in this act and present their findings, from the U.S. viewpoint, to various world organizations. Many of the latter shipping bodies are dominated by European groups and the various shipping regulations being promulgated favor European currency interests. Hampered by severe underfinancing and lack of staff, the DOT is doing their best to upgrade all U.S. interests in shipping codes that would be fair to their products.

CONSUMER AGE

In the 1960's, the age of "consumerism" came into full maturity. More and more people wanted to know more and more information about what was in a product and how much it really weighed. One of the first packaging activities in the early 1960's that was directly influenced by the consumer movement was weights and measures. As far back as Roman times, the concept of weights and measures has been intrinsic to good packaging. Local statutes controlled the various weights and measures on packaging in the states. The net result was total confusion, since laws differed between various states considerably. An attempt to coordinate these state laws was made by the National Bureau of Standards (NBS). In 1965, a model law was drafted by the NBS, but it was really too late. All provisions in state laws that "are less stringent or require different information from" the Labeling Act of 1966 are now invalid. This was due to the passage of the Fair Packaging and Labeling Act (FPLA) in 1966.

FAIR PACKAGING AND LABELING ACT

Under the provisions of the FPLA of 1966 (effective July 1, 1967), responsibility was given to the FDA for the administration of the law. The part of the FPLA dealing with weights and measures is different from the NBS's model law of 1965 and is obviously binding in all states. In addition, the Department of Commerce (DOC) was given the responsibility of helping industry eliminate superfluous packaging. This is under the "proliferation of sizes" statement. The DOC did succeed in eliminating the production of one travel sized toothpast as well as examining the various size packages commonly used for potato chips. The vast importance of such acts cannot be overemphasized to the packaging industry. Basic to this provision is the question as to whether government has the right to dictate the size of packages to be sold in the marketplace.

In recent years, there have been more laws introduced relative to cosmetics, toiletries and nutritional labeling of foods. Various consumer groups have also demanded a new FPLA which would improve upon the provisions of the 1966 act.

With the tightening of money in the U.S., states and local municipalities have started to carefully examine both their overall profits as well as garbage disposability. Even though packaging wastes have been carefully documented to be a very minor part of the waste problem, many localities have focused in on packaging as the prime culprit in their battle for proper waste disposal. Over 1000 bills have been introduced, on various levels, to "ban packaging." To date, only seven on the state level have passed. But future activity is unknown.

In legislation relative to shipping, more careful coordination between the rail, air and motor freight bodies is essential. This has been the situation in recent years and activitv will probably increase.

FUTURE

Obviously, legislative control packaging has only begun and is not finished. Because of the combination of shrinking resources and competition from the emerging Third World nations, shortages of many packaging materials appear to be imminent.

For example, over 209 million short tons of paper and paperboard were produced in 1977. About 175 million short tons were used by the U.S., Canada, Japan and Western Europe. Over the last three years, these areas increased their usage by about 13%. The rest of the world increased

Courtesy of Environmental Protection Agency

FIG. 1.4. VIRGINIA BEACH, VA.
View of Mount Trashmore—an above-grade sanitary landfill built up with municipal solid wastes compacted and covered each day with a six-inch layer of earth. Top soil, when the project is completed, will be added to the slopes and covered with vegetation. In the foreground are maintenance buildings.

their usage by over 50%. There is simply not enough paper fiber available in the developed lands to supply the entire world. While the vast pulp potential available in Southeast Asia, South America and Siberian USSR remain untapped, shortages of pulp appear inevitable in the very near future. Some predictions indicate a pulp shortage in the U.S. by 1981.

There are also finite supplies of petroleum, steel and coal. Glass, whose basic ingredient is unlimited, is made from soda ash. Many western states are opposed to the strip mining techniques used in obtaining soda ash. Also, the severe natural gas shortages in 1978 indicate that the allocation of natural supplies may not be too far away. In packaging, questions such as which materials and how much should be exported may be foremost in the minds of legislators in future years. Coupled with increasing consumerism, future packaging legislation appears to be an inevitable factor in human development.

REFERENCES

ANON. 1975. Packaging laws chart. Mod. Pkg. *48,* (12) 18.

ANON. 1976A. Keeping up with government: A checklist of major contact points. Food Eng. *48,* (12) 16.

ANON. 1976B. Packaging from the viewpoint of packages. Verpeckungs Rundschau *27,* (2) 148-151 (in German).

ANON. 1976C. FDA initiates 5 year cyclic review of additives. Food Proc. *37,* (12) 13.

ANON. 1976D. Resource recovery: A perspective. NCRR Solid Waste Management Briefs.

ANON. 1976E. Timesavers. Pkg. Dev. *6,* (4) 29.

ANON. 1977A. Packaging regulations: How it all started. Food Eng. *49,* (9) 110-111, 113.

ANON. 1977B. Laws/Rulings — All in a weeks worth. Mod. Pkg. *50,* (4) 12.

ANON. 1977C. Laws/Rulings — Its the same the whole world over. Mod. Pkg. *50,* (5) 9-10.

BAUMANN, H. 1975. A step in time builds new products, Canner Packer *144,* (12) 20,22.

BAUMANN, H.E. 1976. Cost impact of food regulations. Food Tech. *30,* (8) 50, 52, 54.

COWARD, A. 1975. Energy and packaging. Brewers' Guardian *104,* (2) 12, 29-31, 36.

MCCORMICK, R. 1975. Nutrient labeling in perspective. Food Prod. Dev. *9,* (9) 7.

TRICOMI, E. 1977. Legislative round-up. Packascope *22,* (16) 1-4.

VANDER KUIL, J.A. 1976. One way or multi-way packaging. Neve Verpeckung *29,* (4) 360-362.

WOOLVETT, R.H. 1976. How the government views: Packaging and solid waste. Can Pkg. *29,* (4) 19-21.

2

Federal Food, Drug and Cosmetic Act — 1938 (The Copeland Act)

The *Federal Food, Drug and Cosmetic Act* (FDC Act) was enacted in 1938 and is the basis of all modern food law in the United States. It was an all inclusive regulation for the time and is divided into the following parts:

(A) General Regulations
(B) Food and Food Products
(C) Animal Drugs
(D) Human Drugs
(E) "R"
(F) Biologics
(G) Cosmetics
(H) "R"
(I) "R"
(J) Radiological
(K) "R"
(L) Import Milk
Import Tea
Caustic Poisons

The major changes to the 1938 Act are:

(1) Oleomargarine Act of 1950
(2) Pesticide Chemicals Act of 1954
(3) Food Additives Amendment of 1958
(4) Color Additive Amendment of 1960

The minor Amendments are:

(1) Interpretive (Miller) Amendment of 1948
(2) Factory Inspection Amendment of 1953
(3) Procedural Amendment of 1954

In the above division, "R" signifies "Reserved" which means that the FDA can use this category when they decide or get approval on a new product mix to be covered under the 1938 regulation.

Many revisions and amendments have served to keep the basics of the 1938 regulation still alive. It is for this reason that United States food law is still very strongly statute oriented. It is important to note that the FDC Act is applicable to all states, territories and for interstate commerce. It is not binding on intrastate movement, since the latter is often subject to the law of the individual state involved. Also, while the FDC Act applies to United States territories such as Puerto Rico, it is not equal in treatment to these areas. Products being manufactured in Puerto Rico must be inspected at their port of entry in the continental United States by the FDA. This can lead to problems with the many drug firms now locating manufacturing plants in Puerto Rico. Since these firms often ship to the United States, FDA inspection of each shipment can often prove to be both cumbersome and expensive.

Information about the FDC Act as well as other federal legislation may be located in the *Code of Federal Regulations,* Title 21 (Food and Drugs). This is published annually under 50 titles and Title 21 is the *Food and Drug* portion. Published daily is the *Federal Register* which covers changes and modifications. Under Title 21 changes are found in the FDC Act ruling. Extremely important revisions (such as in 1975) are often published separately and may be obtained as individual supplements of the *Federal Register.* Also available is the *FDA Consumer,* which publishes "state of the art" articles, summaries and is generally considered the trade book of the FDA.

Various industry groups such as the Institute of Food Technologists, Drug Manufacturers Association, etc., often provide "pink" and "grey" sheets to their members guiding them on new rules and proposed regulations. These are advance bulletins and are based on the ruling that the FDA can talk in "general terms" about a proposed regulation to members of a trade group. These bulletins are in no way considered illegal or unethical.

In the commentary section of the *Federal Register* there is usually found a *Preamble* which explains the law and the Commissioner's philosophy behind that specific piece of legislation. This explains the law better than the law itself; however, since the person writing the *Preamble* is not a disinterested party, no perjurative statement is usually found. Again, the *Preamble* forms an additional source of information to the reader.

Each division of the 1938 law will be discussed relative to its effect on packaging.

GENERAL REGULATIONS

In the early 1960's, consumerism took hold in the minds of the American public. Consumer groups were being formed almost on a daily basis and at an ever increasing rate. These groups were very concerned about the workings of the FDA and its relation to industry. In 1962, a report issued by a 16 member citizen's advisory committee, appointed by the former Secretary of Health, Education and Welfare, Abraham Ribicoff, stated, among other recommendations, that they were "especially disturbed about present industry-FDA relationships." The overall climate accelerated toward consumerism in the late 1960's while the passage of the Public Information Act in 1967 also helped the consumer movement. There was now on record a regulation (Public Law 90-23) that provided guidelines for the public availability of the records of Federal departments and agencies.

The trend in the 1960's and early 1970's toward consumerism reached its climax in 1975. On Tuesday, May 27, 1975, the *Federal Register* published a major revision under *Administrative Practices and Procedures* to Part A (General Regulations) of the 1938 FDC Law. A set of regulations had now become law that codified exactly how the FDA would work with people.

In the 1975 revision, various definitions and labeling exclusions were also included. A package was defined as an "End of Product" package, thus excluding a shipping unit from total coverage. Also included were certain products specifically omitted from the various previously passed labeling acts. The Federal Trade Commission (FTC) had been given the responsibility to patrol size proliferation in packaging — but this was now given to a reluctant FDA. But perhaps the greatest importance of the 1975 revision were the sections on "difference of opinion" and "informal presentation of views."

As mentioned previously, consumer groups were in constant battle with the FDA in the 1960-1975 period. They thought that there was direct collusion between the FDA and industry. Now there was recognized the factor of "difference of opinion" among recognized experts in the field. And included was the definition of "interested person." Any citizen who would be adversely affected by an FDA regulation is an "interested person." These factors resulted in a major breakthrough for the consumer and citizen's groups. For now, qualified representatives of these groups could participate in the various FDA hearings and present their viewpoints.

Courtesy of U.S.FDA

FIG. 2.1. FDA MEETING
Monthly ad hoc meetings in Washington bring together consumer representatives and activists and top FDA officials for a freewheeling discussion of topics that often reflect the latest headlines. At the January 1976 meeting, the agenda included a discussion of FDA's ban on the color additive RED No. 2.

Hearings and Petitions

In previous years, an industrial firm filing a petition generally discussed a plan with FDA personnel *prior* to actual filing. Basic formulations and processing methods might have been discussed, on an informal basis, and a specific plan for filing might have been organized with FDA guidance.These meetings caused great consternation among consumer groups. They believed that these were secret meetings in which both industry and FDA made various deals for the firm's products. In the 1975 revision, the FDA reconfirmed the ability of people to have private conversations with FDA personnel without the conversation appearing on public record. Certain additional provisions were included for these meetings. An FDA official was organized to be the contact between the public and government. A record had to be made of these meetings and it had to be summarized in a report which excluded any confidential data. These summaries then became part of a total administrative record which was open and on file for both public and FDA inspection. In addition, the meetings were published in the "Retrospective Calendar" for the week following the actual event.

Also modified were the actual procedures by which the data received by the FDA was classified, applied and protected. Prior to the age of "consumerism," petitioning the FDA was limited solely to interested parties. Now, the 1975 revision opened up the possibility of citizens petitions. An interested citizen has the right to petition an amendment or appeal an FDA ruling. He must be recorded as an interested person and he may also contest any Commissioner's rulings or object and comment on any petition by an industrial firm.

When the opening up of petitions by citizens appeared, many thought that total chaos would result. Visions of all sorts of citizens attending many meetings were on the minds of both the FDA and business. In reality, only the well organized consumer groups have objected to some proposed regulations. If these groups can establish their overall competence, they are then judged on an equal basis with industrial representatives.

An additional inclusion in the 1975 revision was the clause noting that actions could be reopened. An FDA ruling is issued by the Commissioner. If certain parties object to the already published ruling from a technical standpoint, the action may be reopened prior to court action. All administrative procedures *must* be exhausted prior to court action. Should a party take the matter to court without attempting to reopen it administratively, the entire case can be dismissed by the Commissioner. This clause serves to prevent industry from taking court action against the FDA without adequate administrative review. If new data becomes available after a ruling is published, the action can be reopened, with the Commissioner's approval, even 1 to 2 years later. A "Stay of Action" is also included as a possibility while the case is being re-examined.

To meet criticism that the FDA regulations are difficult to understand, the 1975 revision stated that any action taken by the FDA must contain a summary paragraph written in basic English that could be understood by an average citizen.

Types of Hearings

Included in the major 1975 revision to the 1938 *General Regulations* were responses to FDA criticism from the U.S. Supreme Court. The latter believed that there were too many formal hearings held by the FDA and some type of public informal meeting had to be integrated into the activity of the FDA. The net result was the formalization of several different types of hearings. (See Table 2.1.)

TABLE 2.1. F DA HEARINGS

Type	Public Invited	Data
(1) Formal Evidenture Public Hearing	May not participate	Made Public
(2) AD-HOC Groups	Dependent on size	Made Public
(3) Board of Inquiry	Yes — but can not participate	Made Public
(4) Advisory Committee	Yes — special members	*closed* and *opened* sessions

The most formal meeting held by the FDA is the *Formal Evidenture Public Hearing.* This is organized on court procedures and administered by lawyers. Interested parties can attend provided they have the proper technical background and can play a direct role in the decision making process. The public may also attend the meeting; however, they are not allowed to contribute. Upon completion of the proceedings, all evidence presented becomes part of an administrative record which is open to public inspection with a certain exception. All confidential and/or trade data cannot be reproduced *but* may be viewed during the course of the hearing. This clause has served to hamper manufacturers in their presentation of confidential data. Very few firms are willing to allow their competition the opportunity of viewing their costly, hard-earned secrets.

At the end of the hearing, 30 days are allowed for consideration and for the filing of objections. This is an absolute time and no extensions are possible.

The new form of hearing, added in response to Supreme Court criticism, is an informal type of proceeding modeled after the Ad-hoc Committee concept used by the FDA for many years.

Due to limited finances and personnel, the FDA has often requested industry to provide experts on specific problem areas. These people have had traveling expenses paid by the FDA and their advice has been used as guidance toward the issuance of new rulings. While the overall concept of the Ad-hoc Committee still is a viable one, certain provisions are included on their use in the 1975 revision.

If a meeting of two or less outside experts is held with the FDA, there is no need to publicize the meeting. When two or more outside experts are invited by the FDA as an Ad-hoc group, records must now be kept of the meeting. The meeting date must be published in the forthcoming calender and the names of all attendees must be recorded. The critical provision was one that stated that the meeting had to be open to any interested citizen. In essence, this has served to destroy large Ad-hoc groups since the entire concept of committees can now become cumbersome and unworkable.

To facilitate the overall efficiency of the FDA, the *Public Board of Inquiry* was also organized. An interested party *only* involved in the decision and not the general public may request a *Public Board of Inquiry*. This is held by a chairman and a committee of impartial experts nominated by interested parties. Again, as in the other types of hearings, all documents presented become part of public record. They cannot be reproduced but may be viewed.

Formalization, under the 1975 revision, was also given to the *Public Hearing for an Advisory Committee*. This specified both the rules and regulations for a public advisory committee with specific time limitations. Names of all participants had to be published with two year renewable terms for each member. Provisions were made for both open and closed committee meetings. In the closed ones (to be held to a minimum), confidential data may be presented. *Commissioners' Hearings* are also possible under the 1975 revision.

As is evident in all the different hearings discussed previously, the presentation of confidential information by industry to the FDA has become more and more open to public scrutiny. Coupled with the *Public Information Act* which has no statute of limitations on the disclosure of confidential data by the FDA, the definite trend by industry is for less and less confidential data to be presented for FDA inspection.

On Tuesday, January 25, 1977, Part 2, Subject A of Title 21 was published in the *Federal Register*. It covers and formalizes the Administrative Function, Practice and Regulations originally discussed in 1975. Basically similar to the 1975 regulations, it merely adds to it and codifies all that had been omitted in 1975.

Environmental Concerns

With the passing of the Environmental Protection Act, Part 6 of the FDC Act states that an Environmental Impact Statement must be included in any FDA action before it becomes effective. This has caused problems in certain actions that might not need this statement. There are exclusions such as court actions, factory inspections, etc. But the rule is broad in nature and, if fully enforced, can seriously hamper all FDA activities.

FOOD AND FOOD PRODUCTS

All types of federal food regulations are covered under Title 21 (Food and Drugs) in *The Code of Federal Regulations* (Parts 10-129). The sections are divided into parts and subparts. Each part of Title 21 (Parts 10-129) deals with a specific food regulation.

The beginning parts (1-7) deal solely with administrative standards and general statements while parts 8 on 9 deal specifically with color additives.

Color Additives

In 1906, the original U.S. FDC Act provided for a voluntary system of certification of the purity and safety of dyes used in or on foods. Under the 1938 revision, limits were established on the minute amounts of lead, arsenic and other heavy metals that could be formulated in the colors. It also added a provision setting down the exact specifications of each individual color. Manufacturers submitted samples to the FDA for certification. Covered under this act were the coal tar colors, but not the non-aniline colors such as iron oxide, titanium oxide, and naturally occurring colors.

The Color Additives Amendment of 1960 embraces all colors, both natural and synthetic, used in foods or in packaging materials contacting foods. It was designed to replace the outmoded provisions which governed the use of color in foods under the 1938 Act. Tolerance limitations were established when required as were other conditions for safe use. The system of official certification was continued.

In essence, three groups of colors are permitted for use in coloring of packaging materials for food. Certified FDC dyes and alumina lakes are acceptable. FDA approved pigments and FDA sanctioned colors are also permitted. Any colorant that has an impermeable barrier between it and the food is not subject to the Color Additive Amendment. The colorant may be nonapproved; however, the burden of proof relative to migration, offset, bleeding, etc., rests with the user. All inks should be formulated by using certified or FDA approved colors. It should also use ingredients that are generally recognized as safe or approved materials under the Food Additives Amendment. Lists of acceptable ingredients are available in the various periodical editions of the *Federal Register*.

At present, the entire situation relative to color additives is under close scrutiny by FDA personnel. In the area of certification, the FDA has stated their desire to completely discard their certification program and consider the colors involved to be additives and thus be tested accordingly. There have been many consumer groups urging the above because of the various problems associated with certified colors. It should be noted that if the FDA is successful in dropping certified colors, this will probably also apply to their "batch" certification program. In the latter, the FDA examines the process and controls involved in manufacturing the color and certifies *only* that specific batch.

Foods — General

In the 1938 FDC Act, there are specific prohibitions against the manufacture, introduction and receipt of adulterated foods by *retailers* as well as others in the food chain. (See Table 2.2.) The interstate movement of adulterated foods is also prohibited. An important addition and significant improvement over the 1906 Act was the clause giving the FDA the power to enter a manufacturing plant at "any reasonable time." This is being hotly debated in the courts, but as of this time the FDA still retains their authority.

The removal or information changing/blocking on labels is also prohibited by both manufacturers and dealers. Federal courts are defined and the FDA is given the power to obtain an injunction to stop a manufacturer from producing his product. All foods found to be in violation of the law can be seized. The FDA is also given the alternate power to issue a warning instead of seizure if the problem does not violate public safety.

No standards of identity are mentioned for fresh fruits, dried fruits and vegetables. The power to regulate these products is given to the USDA who in turn created state-federal inspectors which are USDA trained. This is now the situation with the above products. Exempted from this program are avocados, citrus fruits and melons. The 1938 regulation mentions that any standards of identity on file must take into account varietal differences between foods. Different standards are written for these cases. An example is the standards for Italian Style California tomatoes and Rutgers (N.J.) type tomatoes. Both are different and the standards of identity for these two products do reflect their variations.

Packaging is somewhat covered under an interesting adulteration statement. Containers are prohibited which contain a poisonous component (the 1958 Amendment deals more extensively with this matter). Adulteration is defined to include anything which serves to increase bulk or alter the appearance of the product unduly. This has been interpreted by the FDA as including cases of reported overpackaging. In 1940, the FDA acted against a manufacturer of fragile chocolate wafers. Stating that the single layer package had a 75-25% package to mint ratio, the FDA went into court and promptly lost their case. The manufacturer proved that the packaging was needed to prevent product damage. Since then, the FDA has shield away from this area.

TABLE 2.2. GLOSSARY OF ADDITIVES

ADDITIVE	FUNCTION
Acetic Acid	Flavor, Preservative
Adipic Acid	Flavor, Preservative
Agar[1]	Humectant, Texturizer
Ammoniated Glycyrrhizin	Flavor (added to licorice and root beer)
Annatto[1]	Vegetable Color (yellow)
Ascorbic Acid[1]	Nutrient (Vitamin C), Flavor, Antioxidant
Benzoyl Peroxide	Bleaching Agent
BHA: Butylated Hydroxy Anisole[2]	Antioxidant, preserve fat-soluble Vitamins
BHT: Butylated Hydroxy Toluene[2]	Antioxidant, preserve fat-soluble Vitamins
Butyric Acid[1]	Flavor, Preservative
Caffeine[1]	Flavor
Calcium Carbonate	Firming Agent, Color
Calcium Chloride	Firming Agent
Calcium Oxide	Nutrient (calcium)
Calcium Phosphate[1]	Sequestrant, Emulsifier, Nutrient (calcium)
Calcium Propionate	Preservative
Calcium Stearage	Anticaking Agent
Calcium Sulfate	Firming Agent, Nutrient (calcium), Yeast food
Carotene[1]	Vegetable Color (yellow-orange)
Carrageenan[1]	Stabilizer, Texturizer, Emulsifier
Cellulose Gum[1]	Stabilizer, Texturizer
Chlorine	Bleaching Agent
Chlorine Dioxide	Bleaching and Maturing Agent
Chlorophyll[1]	Vegetable Color (green)
Citric Acid[1]	Antioxidant, Sequestrant, Flavor, Preservative
Cochineal[1]	Vegetable Color (red)
Cornstarch[1]	Anticaking Agent
Cyanocobalimin[1]	Nutrient (Vitamin B_{12})
Dextrin	Thickener; prevents sugar from crystallizing
Dextrose[1]	Sweetener (corn sugar)
Disodium Guanylate (GMP)	Flavor Enhancer
Disodium Inosinate (IMP)	Flavor Enhancer
EDTA (Ethylenediamine Tetra Acidic Acid)	Sequestrant Antioxidant
Erythorbic Acid	Antioxidant
Ferric Phosphate	Nutrient (iron)
Ferric Pyrophosphate	Nutrient (iron)
Ferrous Sulfate	Nutrient (iron)
Fumaric Acid	Flavor, Preservative
Glycerine	Humectant
Guar Gum	Humectant, Texturizer
Gum Tragacanth	Stabilizer, Texturizer
Heptylparaben	Preservative
Iodate[1]	Bleaching Agent, Nutrient
Iodine[1]	Bleaching Agent, Nutrient
*Lactic Acid	Flavor, Preservative

TABLE 2.2. *(CONTINUED)*

ADDITIVE	FUNCTION
*Lactose	Sweetener
*Lecithin	Emulsifier, Stabilizer, Texturizer
Magnesium Stearate	Stabilizer, Texturizer
*Malic Acid	Flavor, Preservative
Maltol	Flavor Enhancer
Methyl Cellulose	Stabilizer, Texturizer
Methylparaben	Preservative
*Mono- and Di-Glycerides	Emulsifier, Stabilizer
Monosodium Glutamate (MSG)	Flavor Enhancer
Niacin	Nutrient (Vitamin B_3)
Niacinamide	Nutrient (Vitamin B_3)
Nitrate	Curing Agents, Preservatives, Flavors
Nitrite	
Nitrosyl Chloride	Bleaching Agent
Octyl Gallate	Antioxidant
*Pectin	Texturizer, Creates Gel
Phosphate Salts	Anticaking Agents, Antioxidants, Sequestrants
Polysorbate 60	Preservative
Potassium Acid Tartrate	Chemical Leavening Agent
Potassium Bromide	Bleaching Agent, Maturing Agent
Potassium Chloride	Salt Substitute
*Potassium Phosphate	Sequestrant, Anticaking Agent
Potassium Sorbate	Preservative
Propionic Acid	Preservative
Propylene Glycol	Emulsifier, Humectant
Monostearate	Flavor Enhancer
Propyl Gallate	Antioxidant
Propylparaben	Preservative
Pyridoxine Hydrochloride	Nutrient (Vitamin B_6)
Riboflavin	Nutrient (Vitamin B_2)
Riboflavin-5-Phosphate	
**Saccharin	Artificial Sweetener
Silicates	Anticaking Agents
*Sodium Ascorbate	Antitoxidant
Sodium Benzoate (also called Benzoate of Sada)	Preservative
Sodium Bisulfite	Bleaching Agent Preservative
*Sodium Caseinate	Emulsifier, Binding Agent
Sodium Chloride (table salt)	Flavor, Preservative
Sodium Erythorbate	Preservative, Antioxidant, Color (pink)
*Sodium Phosphate	Sequestrant, Anticaking Agent, Nutrient (phosphorous)
Sodium Propionate	Preservative
Sorbic Acid	Preservative
Sorbitan Monostearate	Emulsifier, Texturizer, Humectant
Stannous Chloride	Antioxidant
Sulfur Dioxide	Bleaching Agent, Preservative

TABLE 2.2. *(CONTINUED)*

ADDITIVE	FUNCTION
Tannin — Tannic Acid[1]	Flavor
Tartaric Acid[1] (Cream of Tartar)	Flavor, Chemical Leavening Agent, Preservative
Tocopherols[1]	Nutrient (Vitamin E), Antioxidant
Thiamin Hydrochloride	Nutrient (Vitamine B_1)
Thiamin Mononitrate[1]	Nutrient (Vitamin B_1)
Vanillin	Flavor
Vegetable Gums[1]	Stabilizer, Texturizer
Vitamin A Acetate[1]	Nutrient
Vitamin A Palmitate[1]	Nutrient
Vitamin D_2[1]	Nutrient
Vitamin D_3[1]	Nutrient
Whey (Dried)[1]	Texturizer
Xanthan Gum	Substitute for gluten (wheat protein) in flour products

[1]Naturally found in some foods.
[2]Controversial — Presently being studied by the FDA

Specifically prohibited are non-nutritive components in confectionery products. This meant that items such as lollipops could not be made since the stick in the lollipop would be in violation of the law. In order to avoid this problem, the term "unless it is functional" has now been added to the prohibition. The use of more than 0.5% alcohol in foods is also prohibited. Below that amount, the alcohol acts as a preservative and if over, the product is subject to the regulations of the Secretary of the Treasury (Federal Alcohol Administration Act) and not the FDA.

Misbranding is prohibited and non-compliance with a published standard of identity is considered to be mislabeling. The areas of labeling and label information are covered in great detail under the FPLA (to be discussed in later chapters). But even in that newly enacted law, consumer groups are now lobbying for improved legislation on percentage labeling as a replacement for predominance of ingredient labeling.

Various additions have been proposed such as the statement that if a product contains over 33 percent sugar (as in some breakfast cereals), it must be noted on the package. This was not included in the 1938 law but will probably be in a future law.

In Parts 10 and 11, foods which do not have standards of identity are specifically covered. Regulations are discussed on how to organize standards for these foods. Colors and pesticides (covered under the Pesticide Chemicals Act — 1954) are excluded. Parts 14-85 mention standards of identity for some specific products. They are listed by item and it is specified that the FDA can approve optional ingredients in standard of identity products. This is rarely done and in fact, the FDA would like to remove all standards of identity in future years.

The Food and Drug Administration has issued new labeling rules to help consumers know just what they are buying when they shop for ice cream. The agency also recommended, but did not order, that ice cream makers let the public know when food dyes or other artificial colors are used. Under the new rules, there could be three types of ice cream labels.

Courtesy of FDA

FIG. 2.2. INSPECTOR WITH CHEESE
A district import inspector takes a sample of imported cheese.

Peach ice cream without artificial flavoring would be called simply "peach ice cream." If an artificial flavor is added — but the ice cream is still more real than imitation — the label would read "peach-flavored ice cream." But if the ice cream is more imitation than real, the label would read "artificial peach ice cream."

The FDA first proposed revising labeling and ingredient standards for ice cream and other frozen desserts in July 1974. Its order is final, except that interested parties could still delay sections of it by filing proper objections.

Ice cream makers may begin using the labeling on June 13 and the regulations will become mandatory on July 1, 1979. The rules also cover sherbet and water ices, and define the difference between ice cream and frozen custard. (See Table 2.3 for other controlled products.)

Part 80 notes that any dietary benefits must be noted in detail on the package. Vitamin data is needed and the specific percentage of the total must also be included. This is only required if a dietary or health claim is made on the product.

The packaging of all foods containing saccharin must bear a label beginning Feb. 23, 1978, stating, "Use of this product may be hazardous to your health. This product contains saccharin which has been determined to cause cancer in laboratory animals."

The labeling is required by the Saccharin Study and Labeling Act which President Carter signed into law Nov. 23, 1977. The act delays for 18 months any move by the FDA to remove saccharin from the market.

TABLE 2.3. PRODUCTS WITH STANDARDS UNDER THE FOOD, DRUG, AND COSMETIC ACT

(1) Cacao products
(2) Cereal flours and related products
(3) Macaroni and noodle products
(4) Bakery products
(5) Milk and cream
(6) Cheeses, processed cheeses, cheese spreads, and related foods
(7) Frozen desserts
(8) Food flavorings
(9) Dressings for food
(10) Canned fruits and fruit juices
(11) Fruit butters, fruit jellies, fruit preserves, and related products
(12) Nonalcoholic beverages
(13) Shellfish
(14) Fish
(15) Eggs and egg products
(16) Oleomargarine, margarine
(17) Nut products
(18) Canned vegetables
(19) Tomato products

Parts 100 and 122 state that artificial coloring must be noted on a package. Butter was specifically exempted because of great pressure from the Wisconsin dairy lobby. For many years, it was illegal to sell colored margarine. In order to evade this regulation, a package utilizing a squeezable bag and a gelatin dye capsule was devised. By breaking the capsule and inserting it in the bag containing margarine, a yellow colored margarine resulted. This was sold all over the U.S., except in Wisconsin. It eventually passed in Wisconsin but even today margarine has more restrictive regulations than butter. Restaurants cannot serve square pats of margarine; they must be cut triangular which is difficult using standard machinery. Also, a sign must state that the establishment serves margarine and it must be in a prominent position. (See Table 2.4 for relevant statutes.) Also, under the Pesticide addition to the 1938 FDC, the Commissioner is given the right to establish regulations which ban or exempt pesticides used on products intended for human consumption. He can also certify the pesticides.

TABLE 2.4. SUMMARY OF STATUTES RELATING TO OLEOMARGARINE

(A) Act of 1902 (32 Stat. 193; 21 U.S. Code Sec. 25): Oleomargarine, butterine, imitation butter brought into a state is to be treated as though it had been manufactured within the state.

(B) Oleomargarine Act of 1950 (Public Law 459 - 81st Congress - March 16, 1950; 64 Stat. 20).

  (1) Amended section 15 of the Federal Trade Commission Act to define oleomargarine of margarine for the purposes of the FTC and also section 407 of the FD&C Act.

  (2) Amended section 230 of the Internal Revenue Code in relation to tax on oleomargarine and declared Congressional policy regarding sales of colored oleomargarine — that such sales depress sales of butter — irrespective of whether the colored oleomargarine is from interstate commerce of manufactured within the state it is sold. (21 U.S. Code 347a)

  (3) Declared that nothing in Public Law 459 shall be construed to permit possession, sale, or serving of colored oleomargarine in a state whose laws forbid it. (21 U.S. Code 347b)

  (4) Added Sec. 301 (m) to FD&C Act: Prohibits sale of colored margarine except under conditions in Sec. 407.

  (5) Provided Sec. 402 (c): Oleomargarine containing or made in whole or in part from filthy, rutrid, or decomposed substance is adulterated.

  (6) Provided now sections to the FD&C Act as given below.

(C) Sec. 407 (a): Colored oleomargarine sold in same state as manufactured shall be subject to the act as if it had been in interstate commerce.

(D) Sec. 407 (b): Provisions under which colored oleomargarine is to be packaged and labeled for sale.

(E) Sec. 407 (c): Notification of customers and identification to be given colored oleomargarine when served in public eating places.

(F) Sec. 407 (d): Colored oleomargarine does not have to be labeled when served with meals

(G) 407 (e): Color standard for colored oleomargarine.

Petition Clearance

Clearance of any specific product remains the responsibility of the organization wishing to manufacture the product.

The first step is for the firm to apply for registration to the FDA. This is done before the material is marketed and even then, the FDA has been known to allow test permits for experimental reasons. A petition is then drawn up and the test methods used as well as safety studies are included. Extractibles must be cleared by the FDA and unless no history is available, it must be provided. In general, dogs and cats have been used as test animals. This is now an expensive procedure and there is talk about using micro-organisms instead of the test animals. The toxicity of materials can thus be judged on a much more economical basis.

The petition must include:

(1) Name, chemical identity and composition of the product
(2) Amount, frequency and time of application
(3) Safety studies
(4) Residue tests
(5) Practical methods of residue removal
(6) Proposed tolerances
(7) Reason for support of petition — are additives needed and why needed?

If residues are found on fruits and vegetables, the petition must show that they can be removed. Tolerances must be proposed and the reasoning behind the tolerances must be included.

A response to the petition should be available in 30-90 days; however, one rarely obtains a decision in that time. There is, in actuality, no average time since the reply depends on the complexity of the product involved. If the FDA desires, they can submit the petition to an advisory committee. This body has 60 days to prepare a report with 30 day extensions possible. Generally, advisory committees keep on schedule and a reply is entirely possible within the specified time allowed. The *Department of Health, Education and Welfare* publishes the proposal or reply in the *Federal Register.* This may be as a restriction, clearance, or complete product ban. If the product is banned, the objector has 30 days to file a protest. Extensions are unlikely to the 30 day allowance. In addition, the objector may request a public hearing and the FDA must agree to hold the hearing. At the hearing, the National Academy of Sciences provides one participant while others may also attend. After the hearing is concluded, the decision rendered is based *solely* on testimony presented at the public hearing and not on all the data available to the FDA either on a confidential or informal basis. As a last resort, the objector can still take the matter to court for judicial review. A 60 day time limit is allowed for court action. And if the matter is a constitutional one, it can even go to the U.S. Supreme Court as a very last resort.

Parts 90 and 128 (a & b)

Called the "Emergency Permit Control," Part 90 is a controversial regulation which allows the FDA to take over a manufacturing plant if the plant demonstrates improper packaging of processed foods which would be injurious to health.

In 1973, the FDA conducted a survey of 400 canneries for low acid foods. The study was aimed at thermal processors and the results were rather startling. About 20% of the processors surveyed did conduct their processing in a manner which was against public health standards. Over 30% were in the borderline area. This startled the FDA and resulted in Part 90 giving the FDA Commissioner the ability to also take over a plant if it fails to register, file, or otherwise not comply with the mandatory items of Part 128b.

In Part 128, it is stated that food processors must register their process and specify their package. The latter means the *package* and not the *material* (an important point). The entire process must be cleared by the FDA and there are certain controls stipulated on thermal processors. Record keeping is stressed and retort operators must be trained in schools approved by the FDA.

Part 128a deals with smoked fish and is essentially a Good Manufacturing Practice (GMP) on how to handle the smoking and preservation of fish. Part 128b deals *only* with frozen shrimp. While frozen shrimp is the largest volume fish involved, it is likely that this part will include all processed fish in later years.

Food Additives Amendment

In 1958, the Food Additives Amendment to the Food, Drug and Cosmetic Act was passed. The rules and regulations proposed by the FDA under this amendment are of significant concern to the packaging industry.

The most important amendment ever made to the original law as far as the food packaging industry is concerned is the Food Additives Amendment of 1958. This law deals with food supply in the United States and takes cognizance of all chemical components of this supply, whether these components enter the food by direct addition or by indirect means, for example, by virtue of migration from a food packaging material.

The packaging industry is concerned with the part of this 1958 amendment that deals with indirect food additives; those substances which can become a component of the food by migration from a packaging material.

The FDA set up the rules and regulations for administering the new law. These rules and regulations included Subparts A through G. The Subparts had the following definitions:

- Subpart A — Definitions and Procedural and Interpretative Regulations.
- Subpart B — Exemption of Certain Food Additives from the Requirement of Tolerances.
- Subpart C — Food Additives Permitted in Feed and Drinking Water of Animals or for the Treatment of Food Producing Animals.
- Subpart D — Food Additives Permitted in Food for Human Consumption.
- Subpart E — Substances for Which Prior Sanctions Have Been Granted.
- Subpart F — Food Additives Resulting from Contact with Containers of Equipment and Food Additives Otherwise Affecting Food.
- Subpart G — Radiation and Radiation Sources Intended for Use in the Production, Processing, and Handling of Food.

If each Subpart is evaluated separately, the full scope of the law is obtained. These Subparts are under Part 121 (Food Additives) of Title 21 (Food and Drugs). The prefix of 121 for each section refers to Part 121.

Subpart A contains several sections and relates primarily to administrative matters pertinent to the administration of the new law. This includes the rules and regulations for the filing of petitions, objecting to proposed regulations, public hearings, judicial review, matters of evidence, etc. This Subpart makes up the substantive law upon which the Food Additives Amendment is administered.

Subpart B is quite short and includes only two sections, but nevertheless, affects the food packaging industry to a great extent. Under this Subpart, Section 121.101 appears and lists all those "Substances that are generally recognized as safe." These substances have now come to be known as having a GRAS status. The forerunner of this section was the so-called "White List." The FDA does not add new materials to this list, but often removes materials from the list when evidence comes to their attention to show that the generally recognized as safe status should not be continued.

Subpart C includes approximately 75 sections and relates to "Food Additives Permitted in Food and Drinking Water of Animals or for the Treatment of Food-Producing Animals." It is not of great relevance to

Courtesy of FDA

FIG. 2.3. TEAMWORK
Inspection team consisting of USDA inspector, FDA microbiologist and FDA inspector are checking the temperature of storage facilities on a recording thermometer.

packaging suppliers because here the food additive is a direct additive, a deliberately added food component. Pet foods packaging materials would fall under Subpart F for indirect additives.

Subpart D relates to "Food Additives Permitted in Food of Human Consumption." This Subpart includes approximately 186 sections and lists those additives which are made directly into food. It is of value to those concerned directly with food processing; however, packagers must know which substances are permitted in a packaging material.

Subpart E relates to substances for which prior sanctions have been granted. These substances are listed in the only section under Subpart E, i.e., 121.2001, which is entitled "Substances Employed In The Manufacture of Food Packaging Materials." Prior sanctioned materials listed in this section means materials that were sanctioned by FDA for use in food packaging materials prior to 1958. Dr. Lehman of the FDA is well-known for his issuance of prior sanctions prior to 1958. No further additions have been made to the prior sanctions list since 1958.

Subpart F relates to "Food Additives Resulting From Contact With Containers of Equipment and Food Additives Otherwise Affecting

Food." This is the most relevant Subpart for the packaging industry. The Subpart contains approximately 97 sections. Within these sections many components of food packages are regulated. Section 121.2501 relates to Olefin polymers which includes polyethylene, polypropylene, etc. Section 121.2502 relates to nylon resins. Even though there is no regulation for a food packaging material such as polyethylene and a determination can be made by simply referring to these regulations, it is also necessary to obtain assurance from a particular supplier that the material in question actually meets with provisions of a specific regulation. Polyethylenes are not all the same and many polyethylenes are not food grade. Therefore, even though there is a regulation under Subpart F pertaining to polyethylenes, permission must be obtained, in writing, to offer assurance from suppliers of polyethylene that the grade of polyethylene to be used by a supplier in a good packaging application is "food grade" by virtue of its compliance with Section 121.2501 of regulations under Subpart F. This holds true for all other packaging components, such as lubricants, plasticizers, coatings, adhesives, etc.

Extraction Tests.— Several FDA regulations under Subpart F require that extraction tests be run. This is required by some regulations to make sure that the chemicals involved do not migrate to the food beyond the level established by FDA to be safe. Analytical tests can be performed via extraction to make sure products do not extract beyond the safe limit.

Extraction data may be complex. All extraction data must contain the amount of a single substance which migrates under standard conditions and the total contamination of the food product. In addition, toxicity data must be available for FDA evaluation.

Migration of additive is dependent on time, temperature, area, and type of food. While extraction studies may be conducted on the specific food, most tests utilize special extracting solutions. This is due to the wide range of variables inherent in food products which cause problems in obtaining accurate results. For particularly sensitive applications radioactive tracer analysis is used. When tagged carbons are employed, excellent results are obtained. However, radioactive tracer tests are expensive and require extremely intricate equipment. Due to these factors, the process is not in widespread use.

In order to formulate a range of solvent solutions capable of simulating food extraction, several factors are significant. The various solution types that should be included are acid, alkaline, and neutral solutions as well as alcoholic and oil materials. A representative list suggested by the United States Food and Drug Administration is: (1) distilled water; (2) 3% aqueous NaCl; (3) 3% aqueous $NaHCO_3$; (4) 3% aqueous acetic acid; (5) 3% aqueous lactic acid; (6) 20% aqueous sucrose; and (7) lard or vegetable oil.

Courtesy of FDA

FIG. 2.1. PLANT CHECKS
FDA inspector checking confectionery production.

The list of solutions recommended by the French authorities is more complex: (1) distilled water; (2) 10% ethyl alcohol; (3) 50% ethyl alcohol; (4) 95% ethyl alcohol; (5) 10% acetic acid; (6) 1% NaCl solution; (7) 5% NaCl solution; (8) 10% sucrose solution; (9) 2% citric acid solution (pH=4.5); (10) lard; (11) arachide oil.

It is essential that all extractants be used correctly to arrive at accurate determination of the migrant. Since various migrants often require different extracting solutions, the proper test method for the specific chemical must be used. For example, highly poisonous substances require extremely sensitive test solutions.

Migration studies should actually be conducted for the entire shelf-life of a packaged food. In principle, the duration involved makes long-term tests impractical. Most tests are conducted at accelerated temperatures and the results correlated with actual conditions. United States authorities state that the test conditions used must correlate to actual packaging conditions. If a film is autoclaved during a packaging operation, migration tests must be conducted at the autoclaving temperature. Other nations feel that the test conditions must be more severe than actual practice. Since it is often impossible to know the total packaging cycle, extremely stringent test conditions are usually employed.

The diffusion of migratory substances into food products increases with increasing temperatures. It has been reported that extraction is fairly uniform for the first 5 days at a temperature of 60°C (140°F). After the fifth day, the rate decreases and extraction ceases by the tenth day.

The thickness of the sample used should be uniform. A 1.0 mil sample is recommended since it encompasses most commonly used plastic films and sheets. Total migration should be expressed parts per million, and a ratio of 0.3 ml solvent/cm^2 of surface is preferable for the test method.

After allowing the sample to be exposed to the solvents chosen at the specified time and temperature cycles, quantitative identification of migratory compounds is made by gas chromatography or conventional analytical methods.

Subpart G relates to radiation and radiation sources in the handling of food.

Subpart F is the only Subpart which contains regulations which require compliance. If a material can be added directly to food, is GRAS, or has been prior sanctioned, it is not regulated as an indirect food additive. Accordingly, the food packaging industry is concerned with those violations of FDA regulations which cover indirect food additives.

Food Additive Definition.— The definition of a food additive is of great value in fully understanding the law. The 1958 Amendment provides that a food shall be deemed to be adulterated if it contains a food additive. Once a food contains a food additive or if a food packaging material is capable of giving rise to a food additive, penalty provisions of the act apply. The definition of a food additive is most important. A food additive is not any substance added to food directly or indirectly.

The Food Additives Amendment as passed by Congress included a special definition for a food additive. The legal definition of a food additive means "any substance the intended use of which results or may reasonably be expected to result, directly or indirectly, in its becoming a component or otherwise affecting the characteristics of any food (including any substance intended for use in producing, manufacturing, packing, processing, preparing, treating, packaging, transporting, or holding food; and including any source of radiation intended for any such use), if such substance is not generally recognized, among experts qualified by scientific training and experience to evaluate its safety, as having been adequately shown through scientific procedures (or in the case of a substance used in food prior to January 1, 1958, through either scientific procedures or experience based on common use in food) to be safe under the conditions of its intended use." This term (food additive) does not include pesticide chemicals, or any substance used in accordance with a sanction or approved of before this enactment.

A food additive is actually any substance that can become a component of food by direct or indirect means unless that additive is known to be safe or is prior sanctioned. If that additive is known to be safe, such as GRAS materials, or is prior sanctioned, then that substance is not a food additive. In other words, all substances that can become a component of food are food additives unless declared otherwise by FDA. The FDA has declared some substances nonfood additives by listing them on the GRAS list, and has declared other substances nonfood additives by prior sanctioning. These substances appear in Subpart B and E, respectively.

All other materials of the food industry which are used both in food directly and in packaging materials for foods are treated by the FDA in a separate manner. The FDA requires a petition to be filed by the chemical manufacturer requesting a permit to use a product in food or in a food packaging material. After FDA reviews the toxicological information in the petition and finds it satisfactory, the FDA will issue a regulation for the safe use of the chemical in question. This regulation then regulates the use of the material. Regulations which cover direct additives for food appear in Subpart D, and regulations which cover indirect additives appear in Subpart F. The regulations appearing in Subpart F are of most relevance.

By the issuance of a regulation for a particular substance, FDA has declared that substance safe and, therefore, not a food additive so long as the use of the substance complies with provisions of the issued regulation. A regulation states that migration of certain substances from a coating shall not exceed 0.5 mg/in.^{2} of contact surface in the presence of a certain solvent under particular extraction conditions. If, in fact, the extraction level from this regulated coating under these conditions is higher than this amount, then the packaging material in question would be capable of introducing a food additive into the food supply. This occurs because the packaging material components could enter the food supply at a level beyond limits known to be safe by FDA. At this point, violation of the 1958 Food Additives Amendment exists and all the provisions of the Food, Drug and Cosmetic Act apply, including the seizure provision.

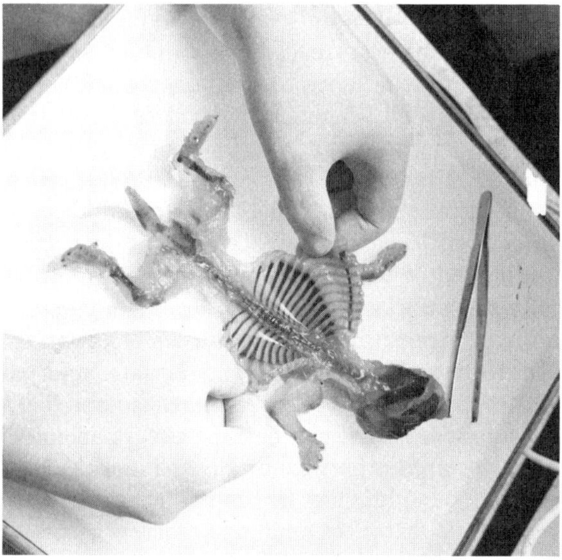

FIG. 2.5. TERATOLOGY Examination of organism to determine malformations due to food / drug interaction.

Courtesy of Food and Drug Research Laboratories, Inc.

When the Food Additives Amendment was passed, a simple test for clearance cost about \$10,000. Animal tests cost about \$50,000, while full scale generational testing cost \$150,000. Presently, these tests can amount to over \$350,000 and the expense is now prohibitive for many products. It has often been suggested that this can lead to a strangulating effect on many materials causing their elimination from the market.

The FDA is also not satisfied with GRAS substances and since 1975 has been rescinding these materials. All chemicals used in foods and packaging materials are scheduled to be tested by FDA investigators in future years. These studies are planned to include all previously approved materials — even simple plastics such as polyethylene. More and more data is needed on every material and the FDA will then review their approvals every 5 years with any problem plastics scheduled for review every 2 years. The entire FDA program is both time-consuming and comprehensive.

THE PLASTIC SOFT DRINK BOTTLE — STATUS (1977)

The war of the plastic bottles — acrylonitrile against polyester — is not completely over even though a ban on acrylonitrile bottles became effective in December 1977.

Coco-Cola bottlers are making a switch from the quart-sized acrylonitrile bottles produced by the Monsanto Company. These were being marketed heavily in the Northeast until March 1977, when the ban was first proposed by the United States Food and Drug Administration.

Starting in Spartanburg, S.C., Coco-Cola bottlers in the Southeast have been introducing two-liter (half-gallon) bottles of the rival plastic, polyethylene terephthalate, chosen years ago by Pepsico Inc.

Since the nation's two largest makers of soft drinks have settled on polyester for their large bottles, the main polyester supplier, the Goodyear Tire and Rubber Company, is rapidly expanding its shipments to several fabricators of the bottles. These include the Amoco Chemicals Corporation, Sewell Plastics Inc., the Hoover Ball and Bearing Company, Owen-Illinois Inc., and Continental Diversified Industries.

For some time, the Monsanto Company has indicated that it has not abandoned the fight. Under the law by which the FDA reaffirmed its proposed ban, Monsanto had until November 21, 1977, to challenge the ban in a Federal appeals court.

The FDA ruling is based on several considerations. One is that, no matter how carefully the plastic, polymerized acrylonitrile, is made, it will contain stray molecules, called monomers, of acrylonitrile. Under some conditions, such as storage of filled plastic soft-drink bottles in a warehouse heated continually for six months to 120° F, these acrylonitrile monomers could leach out of the plastic and invade the drink.

FDA-ordered tests, conducted by the Dow Chemical Company for the Manufacturing Chemists' Association, showed that laboratory rats fed very large doses of the monomer for their lifetimes developed an excess of tumors above the normal rate. The offspring of pregnant rats fed the monomer showed an excess of birth defects.

If the monomer can leach from the plastic into the soft-drink, it becomes a food additive subject to FDA jurisdiction. But Monsanto has asserted that this could not happen to real bottles under real conditions.

If the bottles, which Monsanto calls "Cycle-Safe," were stored at 48.9°C (120°F), the pressure from the carbon dioxide in the drink would build up from the 26 16/sq in. at the typical refrigerator temperature of 4.4°C (40°F) to 132 16/sq in. At 132 16/sq in., the bottles cease to be functional containers in just a few days, and many would likely rupture within two weeks. At the least, they become enlarged and badly deformed, the liquid level drops noticeably and the beverage becomes flat and loses its familiar flavor.

In other words, there is no way a Cycle-Safe bottle, filled with soft drink, could survive this kind of abuse for one month, let alone six, says Monsanto.

To get around the problem of gas pressure buildup, Monsanto said the FDA simulated beverage-filled bottles by using water containing 3.0% of acetic acid and 8.0% of ethyl alcohol.

A legal challenge on the details of the tests could prove difficult. The judges of the Federal Court of Appeals for the District of Columbia, who have heard many cases on such environmental issues as the risks from nuclear power plants have ruled repeatedly that the courts should not enter technical and scientific disputes except to insure that the issues are argued fully, as required by such laws of the National Environmental Policy Act of 1970.

However, Monsanto lawyers reportedly are considering whether to challenge current administration on the laws governing food additives, which include an absolute ban on anything that causes cancer during so-called appropriate tests, down to the lowest limit of detection.

Monsanto's position paper (July 15, 1977) says that the "underlying principles" of the absolute ban, while not cited directly by the FDA, influenced its decision. But did Congress envision a 20-yr advance in the sensitivity of chemical tests from detection of parts per million to quantities thousands of times smaller?

So far, Monsanto has only won a procedural battle in the courts. In March, 1977, the District of Columbia appeals court suspended the FDA's ban pending a full "evidentiary hearing."

But this hearing did not change the FDA's ruling, and, at any rate, the ban was in effect. In February, after announcement of the animal test results and indications of FDA action, orders for the bottles from Coca-Cola bottlers dried up and Monsanto closed four manufacturing facilities and laid off 800 workers.

In May, E.I. duPont de Nemours & Company announced health data on some 500 long-time workers at its Camden, S.D. plant that linked exposure to acrylonitrile during the plant's start-up in 1951-53 to an excess of cancer cases and cancer deaths. DuPont announced steps to assure a tenfold reduction in workers' exposure to acrylonitrile.

DRUG, MEDICAL DEVICES AND COSMETICS

Under the 1938 FDC Act, *drugs, devices and cosmetic* regulations are found in parts 130-1500. Animal drug regulations (Subchapter C) are located in parts 130-151, while part 133 specifically contains GMP's for animal products. Human drug regulations (Subchapter D) are found in parts 200-600; biologicals are located in parts 600-680 and cosmetic regulations are in parts 700-730. GMP's for drugs are in parts 201, 207, 211 and 229.

General Considerations

As in food regulations, adulteration of drugs is carefully considered; however, the regulations on adulteration for drugs are much more rigid than those for foods. A drug is considered to be adulterated if it is composed of filthy, putrid or decomposed substances. If the method by which the product is packaged is contaminated, then the drug is also considered to be adulterated. The facility or control method used to package the drug must also conform with all existing GMP's. The drug must not be diluted nor contain an unsafe ingredient. It also is specified that the drug is considered to be adulterated if it does not follow the various standards found in the National Formulary, U.S. Pharmacopeia and Homeopathic Pharmacopeia. In the latter, drugs are categorized and their packaging techniques discussed. The stipulation that the drug must adhere to these non-governmental standards is a unique one since it is an example of an official governmental organization (FDA) administering non-governmental standards. In animal drugs, a feed is adulterated if it contains any drug that is unsafe. (See Table 2.5 for regulation changes.)

TABLE 2.5. DRUG REGULATION CHANGES SINCE 1938

Date	Regulation of Act
1962	New Drug Act (Kefauver-Harris Amendment)
1965	Drug Abuse Control Act
1967	Computerized Data Bank Organized
1970	Comprehensive Drug Abuse Control Act
1971	NDA and IND Guidelines organized with cooperation of Pharmaceutical Manufacturer's Association (PMA)
1972	Biologicals Covered by FDA
1976	Medical Device Amendment
1978	GMP's for Medical Devices

Misbranding occurs if a drug containing a narcotic or hypnotic is not properly labeled with a warning statement. The label must also contain the manufacturers name, address and other required information. This includes the systematic name of the drug. Directions for use must be included to avoid misbranding. These directions are not considered to be complete if special instructions for children, pregnant and lactating women are not included, when applicable. If the item is subject to deterioration, it must be specified on the label.

Certain products such as insulin, antibiotics and color additives must be certified on the "batch" basis. New drugs are registered and antibiotic certification may be waived if after the FDA checks 50 consecutive batches and found all to be acceptable.

There are certain exemptions to the above regulations. Bulk drugs intended for repacking are not included. The FDA is concerned about this exemption and it is felt that in the near future a GMP covering bulk drugs will be issued. Prescription drugs and new drugs for clinical testing are also exempted.

There are 5 basic steps necessary in order to file for a new drug approval with the FDA:

(1) Safety of drug must be proven
(2) Effectiveness of drug must be proven
(3) Composition of drug defined
(4) Manufacturing and packaging methods specified
(5) Samples need to be submitted

After 180 days, the FDA will advise whether the product is approved, rejected, or a hearing will be held. The FDA relies on advisory committees to judge the acceptability of a new drug. There are various committees organized for special product mixes and the committee selected to work with the FDA should have expertise in that area. If the petitioner does not agree with the decision, appeal can be made to a circuit court or the Supreme Court. In 1973, the U.S. Supreme Court, in an important landmark decision, upheld the FDA's responsibility in all areas of drug and device regulations. This has served to strengthen the FDA's role in drug affairs and their responsibility in the safety of public health.

New Drug Act — 1962 (Public Law 87-871).

In 1962, the New Drug Act (Kefauver-Harris Amendments) was passed. The most important aspect of this new regulation was that the FDA was given authority to initiate proceedings to prohibit the sale of a drug cleared for safety as a new drug under the act of 1938 if the agency believed there was a lack of substantial evidence of *efficacy*. A lag period of two years was allowed until October 10, 1964, so that the sponsors and the FDA might assemble any needed information. Faced with clearing and ruling on about 5000 ethical drugs and 100,000-500,000 over-the-counter (OTC) drugs, the FDA immediately organized various advisory committees to aid them in their efforts. The added responsibility of clearing OTC drugs was a considerable one since it had previously been the responsibility of the Federal Trade Commission (FTC). Now, advertising copy as well as drug efficacy for OTC's was in the realm of the FDA.

Under this act, drugs are defined as "articles intended for use in the diagnosis, cure, mitigation, treatment, or prevention of disease in men or other animals; and articles (other than food) intended to affect the structure or any function of the body of man or other animals." It is important to note that this definition includes items other than those normally considered to be drugs by the layman. (See Table 2.6.) It is a rather broad definition and an all-encompassing one.

TABLE 2.6. TYPES OF RECALLS

Class	Name	Example
I	*Serious Recall*	Imminent threat to public health removed from consumer level, i.e., botulism scare in mushrooms
II	*Potential Hazard*	Removed from retail level, i.e., Beef Stew cans — infected
III	*Law Violation*	Removed from wholesale level, i.e., relatively minor law infraction

To discharge its responsibility as to the safety and efficacy of all drugs, the FDA divided drugs into two categories according to the following definitions:

New Drug — any drug marketed after 1962 or one in use at that time which had not previously been cleared by the FDA or generally recognized as safe.

Old Drug — any drug in use before 1962 which is generally recognized as safe.

The distinction between new and old drugs it is not always clear-cut, and it is imperative that a drug company establish whether their product is new or old. This is generally done by informally checking with the FDA.

Regulations promulgated under the 1962 Act require that a New Drug Application (NDA) be filed with the FDA before a drug meeting the new drug definition can be marketed. If a drug is an old drug no NDA is required. A word of caution here — if an old drug is changed in form (such as aspirin into a time-release capsule), or a drastic change in packaging is instituted, the drug could be redefined as a new drug. This would necessitate the filing of an NDA. The FDA's view on a package change is that it may deprive the drug of "its genetic recognition of safety" and makes necessary proof that the proposed package will not be "reactive, additive or adsorptive to an extent that it significantly affects the identity, strength, quality or purity of the drug."

The NDA must provide sufficient data to show that the drug in question will be safe and effective for its intended use. To obtain this data, the drug manufacturer must go through a series of preliminary steps, each of which is closely monitored by the FDA.

The first step, after completing the research on the drug at the drug manufacturer's research laboratory, is to file with the FDA a "Notice of Claimed Investigational Exemption For A New Drug," or IND. The IND sets forth an outline of the clinical studies that will be pursued. The quantitative composition of the drug, the methods used to insure its identity and uniformity, and evidence for safety based on animal investigations must be clearly stated. Clinical studies may proceed after the FDA has reviewed and generally approved the IND, but may be stopped any time the FDA has evidence of probable toxicity.

Other Considerations

The FDA also requires that all drug manufacturers register annually and notify them when changes both in product and label occur. Samples must be forwarded to them and in return the FDA provides the drug firm with a registration number. Exemptions to this regulation include retail pharmacies, physicians and research facilities. Drug firms are also inspected once every two years. A foreign firm, importing a pharmaceutical into the U.S., can also register with the FDA. Their location may be inspected by U.S. FDA personnel or the FDA may elect to contact their foreign counterpart and request them to perform U.S. cleared tests for the item. Foreign manufactured drugs may also be cleared by FDA personnel on a "batch" basis.

FIG. 2.6. DRAIZE EYE TEST

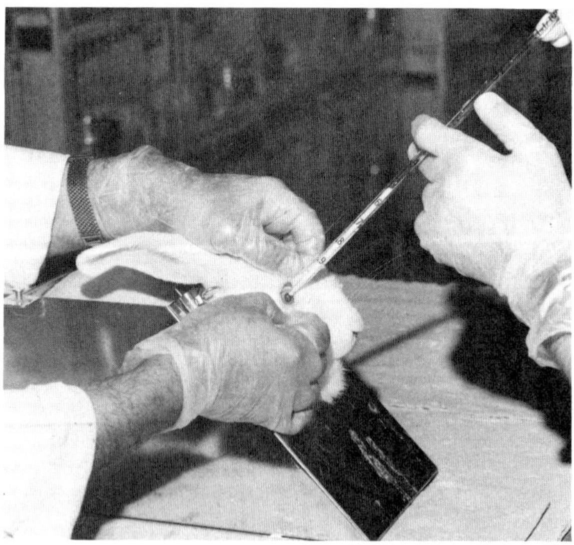

Courtesy of Food and Drug Research Laboratories, Inc.

GMP — Drugs

Newly proposed GMP's have been the recent concern of the FDA. In essence, these GMP's merely codify commonly accepted practice. An important aspect is that biologicals have been formally included in the proposals. Again, this has been the situation since 1972 when the FDA obtained full control over those products from the National Institute of Health (NIH).

The proposed GMP's acknowledge the use of computers in maintaining proper control records. Manual records must also be organized by a responsible employee. The role of quality control in a lab facility is discussed. Questions relative to animal testing, routine lab examinations and overall housekeeping are answered. Employee training and clothing required are included. If a consultant is used by a drug firm, his competence must be demonstrated to the FDA. In the area of package materials, inventory controls and material flow are tightened. Quarantine arrangements are discussed for incoming materials. Labels must be formerly reconciled and fully accounted for in each run. The area used for packaging must be cleared of material after each run. Continuous inspection of package goods must be conducted during the run to ascertain the functionality of the material. All components of the package must be coded. This includes the outer shipper as well as the inner labels. Record keeping is stressed for all package materials.

While the FDA has complete responsibility for both OTC and ethical drugs, the above GMP's apply only to ethical drugs. They have not yet been extended to all drugs and are based on informal agreements with the pharmaceutical manufacturers. It is expected that in the future, the FDA will extend these GMP's to all drugs.

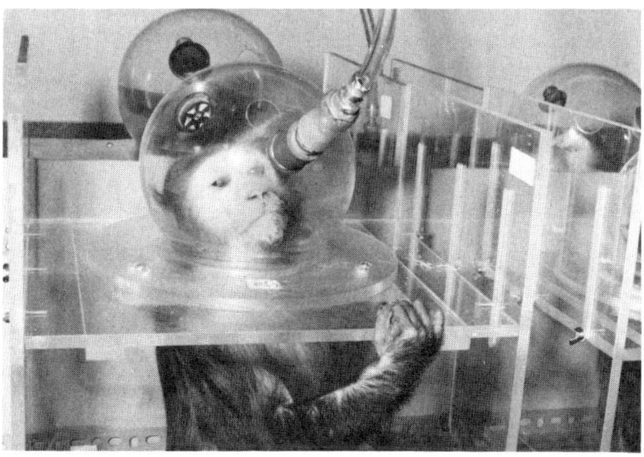

FIG. 2.7. INHALATION TESTING
Animal being tested via inhalation.

Courtesy of Food and Drug Research Laboratories, Inc.

New Drug Law Proposed

In late 1977, the Carter Administration proposed a sweeping legislative overhaul of the nation's pharmaceutical laws. If enacted, the proposal would change virtually every aspect of the Federal regulation of prescription and over-the-counter drugs and impose profound changes on the entire pharmaceutical industry.

The intent of the proposal, as drafted at the direction of the Commissioner of the FDA, is to simplify and strengthen all the drug legislations that were passed this century, including the original 1906 Act and the major amendments adopted in 1938 and 1962. Highlights of the proposal are:

(1) Speeding-up of new drug approval
(2) Easier methods of removing dangerous drugs from market
(3) Fostering competition by easing rules under which companies may market drugs developed by another firm
(4) Forcing proper monitoring of drug side effects
(5) Making test data public
(6) Enacting civil penalties of up to $10,000 for failure to comply

Officially entitled, "Drug Regulation Reform Act of 1978," the draft proposal has 68 sections. One rather interesting section would create a national monitoring and surveillance system for new drugs by requiring that reports on the use of new products be made by doctors and hospitals to the manufacturer. The companies would then be mandated to report side effects to the FDA.

Medical Devices Amendment

Initiated by President Nixon and passed during President Ford's term in 1976, the Medical Device Amendment formally gave the FDA jurisdiction over the control of medical devices. This had always been the responsibility of the FDA; however, it had been spelled out in rather ambiguous terms in the 1938 regulation. The marketing, in 1970, of a useless radiological cure for cancer caused the FDA to remove the product from public sale. A court case followed with the FDA winning. After congressional action and a friendly administration, the new regulation was issued.

The Medical Device Amendment states that a device is a "health care product that does not add to the chemical reaction of the body nor is it metabolized by the body." It gives the FDA specific authority to issue GMP's and also to ban or withdraw a device from the market without having to prove its effectiveness. (See Table 2.7.)

TABLE 2.7. 1977 PROPOSED GMP OUTLINE FOR MEDICAL DEVICES[1]

Section	Title
A	General Provisions and Definitions
B	Organization of Personnel
C	Building Section
D	Use of Equipment for Manufacturing
E	Control Components and Containers
F	Production and Process Controls
G	Packaging and Labeling Control
H	Control and Distribution
I	Product Evaluation
J	Records and Reports

[1]Section L (sterilization) originally in the 1975 suggestion has not been deleted.

About 8000 medical devices are currently on the market. The FDA has categorized these devices into three groups:

(A) General Controls Apply; e.g., *basic types*
(B) Performance Standards Apply; e.g., *diagnostic tools*
(C) Premarketing Approval Necessary; e.g., *pacemakers*

A device manufacturer must give the FDA a 90 day notice of a new product. At the end of this time, the FDA will inform the manufacturer of the category of the product. An industry advisory group will judge the device as to its group classification and usefulness.

New final GMP's for all medical devices were issued in early 1978.

On March 1, 1977, proposed GMP's were published in the *Federal Register*. These were a significant improvement over the GMP's originally suggested in August, 1975. Basically considered to be an umbrella GMP, it is in actuality a two tiered GMP which divides medical devices into *critical devices* and *non-critical devices*.

Critical devices are those devices which act to support or sustain life or are implanted in the body. Failure to perform causes death. If there is a dispute as to the proper category of a device, it is left to an advisory committee for final judgement. All other devices are considered to be non-critical devices. The FDA also proposes to publish subsequent GMP's. These will be two different types of GMP's, i.e., generic classes of devices and other classes. The generic class will include items such as eye-glasses and heart valves. The other class will stress properties such as sterile devices, plastic devices and electrical types. Section L, originally discussed in the 1975 suggestion, has now been deleted.

There are still several areas that are objectionable to industry. One is the FDA statement that each firm manufacturing medical devices maintain a separate quality control department. This must be entirely distinct from the production department. Smaller device manufacturers maintain that this will create an undue hardship on them. The subject of automated quality control processes is an additional point of controversy. An automatic control system must perform exactly what it is intended to do and GMP's are needed on automatic data intake. Also buildings and various environmental factors are discussed. While factors such as ventilation are important for the manufacture of critical devices, they have been somewhat relaxed for the non-critical category. As long as a manufacturer formulates a procedure which appears to be satisfactory, it is acceptable to the FDA. Good GMP's are the responsibility of the manufacturer.

In the original 1975 GMP's, a standard for the ability of suppliers to meet standards of quality was proposed. Industry objected to this requirement and it was omitted. Critical operations were also considered and defined as those operations where failure could cause a serious hazard to those using the device. Those operations must be performed by people with proper backgrounds and education.

The overall intent of the 1977 proposals was to strengthen the area of critical devices and somewhat relax the non-critical device category. Industry and other interested parties have 120 days for comment on these proposals.

Vitamins and Minerals

Basic to FDA philosophy is the concept that vitamins and minerals are unnecessary supplements to a properly balanced diet. And because of this belief as well as problems in the entire area, a new regulation on the above products has been issued January 1, 1978. All minimum potencies must be listed on both vitamins and minerals. Specific limits have been placed on those vitamins (A and D) that are toxic when taken in massive doses. Lower limits on the label have been set for children and lactating/pregnant women. All components must be listed as well as their sources and forms. Any therapeutic claims for dietary supplements will be considered to be misbranding after January 1, 1978. Questionable products will be subject to seizure or injunction proceedings by the FTC within 90 days. If the FTC ignores the product, the FDA can act after the specified time period.

Recalls

There are three different categories of recalls as outlined in Table 2.4. It is critical to realize that all recalls have increased by 33% since 1970.

The basic increase has been due to drug recalls. An example of a recent recall is "Coca-Cola" and "Pepsi-Cola" packaged in acrylonitrile bottles. This is a Class III recall since the FDA said that bottles still on retail shelves may be sold. The FDA order also permits bottles in warehouses to be sold as long as they are not in interstate commerce.

Cosmetics

Unlike drug regulations, a cosmetic manufacturer does not have to prove the efficacy or safety of a product prior to its market introduction. The FDA must prove that the item is unsatisfactory. This has been the situation since the original 1938 law. All essential adulteration statements that apply to drugs also apply to cosmetics.

Cosmetics, as do drugs, must conform to all the regulations of the 1970 Poison Prevention Packaging Act (PPPA). This is particularly true for various coal tar products used in hair preparations. Certain exemptions exist for labels such as very small size packages which are too small to receive proper label copy. It is important to realize that over two-thirds of all complaints received by the FDA relate to hair preparations, eye products and until recently, vaginal deodorants.

In order to give the FDA increased authority in the cosmetic area, a voluntary registration program was set up with the Cosmetic, Toiletry and Fragrance Association (CTFA). The cosmetic industry has complied with this voluntary request and the data assembled by the FDA is put into computers to ascertain which compounds are considered safe and which materials required additional data. These groups are then evaluated by industry experts and appropriate action is taken.

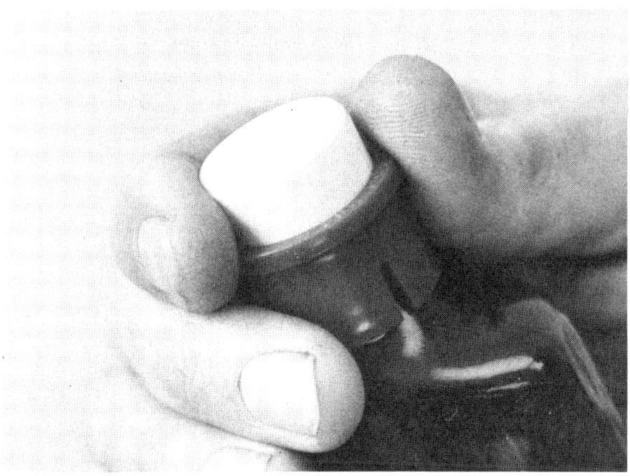

Courtesy of Food and Drug Research Laboratories Inc.

FIG. 2.8. CHILD RESISTANT CLOSURE
Such closures conform to the regulations of the PPPA of 1970.

In aerosol labeling, certain required warnings appear on different products. Hair sprays must contain a warning for eye protection. Inhalation and incineration warnings also appear on the label. There is an irritation warning on feminine deodorants.

Hyperallergenic products must be pretested and determined that the item listed as hyperallergenic causes less allergic reactions than competitive products. This data is obtained by human tests conducted under medical supervision. In an action taken by a manufacturer of hyperallergenic products, the question as to whether the FDA has jurisdiction over this area has been raised. The problem is still in the courts; however, a 2 yr stay of action has been granted in this case.

Ingredient labeling has still not been resolved finally by both the FDA and CTFA. The regulations state that any packages ordered after May, 1976 must have ingredients labeled, but that old packages may be shipped until November, 1976. Any retail stock sold in stores now can be sold indefinitely. The regulations proposed were considered too weak by consumer groups. Table 2.8 illustrates each viewpoint. The first bill introduced in the Carter administration (January 21, 1977) dealt with a proposed revision to the FPLA. Included in this draft are various consumer comments on cosmetic labeling. While ingredient labeling is still voluntary and not yet enforceable by the FDA, most cosmetic manufacturers are meeting the voluntary requirements. In a comprehensive article on the subject, the *New York Times* (November 19, 1975) stated, "Many cosmetic manufacturers see no need to list ingredients and insist that even after it is done consumers won't read the label."

TABLE 2.8. COSMETIC INGREDIENT LABELING

Item	Consumer Complaints	Proposed Regulation
Ingredients listing method	Percent of	Descending order of predominance by weight
Components included	All inclusive	Trade secrets omitted and fragrances/flavors to be unidentified. Listed as such.
Warnings on packages	Agree on warnings	Safety warning obligatory on perfume sprays (0.5 oz) and on metered sprays (2 oz)

REFERENCES

ANON. 1975A. Juice drink report: A survey of FDA nomenclature labeling rules. Dairy and Ice Cream Field *158,* (12) 48, 51.

ANON. 1975B. Washington survival guide. Pkg. Eng. *20,* (5), 36-69.

ANON. 1975C. Change in FDA regulation said based on food supplies. Food & Drug Pkg. *32,* (10) 4, 21.

ANON. 1976. FDA initiates 5-year cyclic review of additives. Food Proc. *37,* (12) 13.

ANON. 1977A. A guide to government packaging regulations: Part 9. Drug and Cosmetic Regulations — II, Pkg. Printing and Diecutting *24,* (9) 24, 26.

ANON. 1977B. Nutrient labeling more important than percentage labeling, CU says. Food Chem. News. *18,* (51) 35-37.

ANON. 1977C. Drained weight labeling. Food Tech. In Australia *29,*(1) 27-29.

ANON. 1977D. Toxicological insignificance bill introduced in congress. Food Chem. News *19,* (9) 27-30.

ANON. 1977E. Recent regulations affecting cosmetics. FDA Consumer *11,* (3) 22-23.

CHAPMAN, S. 1975. New label rules: An interpretation. Candy and Snack Industry *140,* (13) 34-40.

CROCCO, S.C. 1977. FDA's nutrition survey. Food Eng. *49,* (2) 50-52.

DAVIS, R.D. The challenge of ingredient labeling. The Brewer's Digest *50,* (6) 28, 34.

EIERMANN, H.J. 1976. New cosmetic labeling requirements. Assoc. of Food & Drug Officials Quart. Bull. *40,* (2) 73-84.

HICKOX, R. 1976. Canners oppose labeling expense. Food and Drug Pkg. *34,* (11) 1, 45.

HOPKINS, H.C. Helping the consumer play safe. FDA Consumer *9,* (4) 11-12.

NORMAN, C. 1975. Washington report on food labeling. Nutrition Today *10,* (3) 28-30, 34, 36.

SCHRAMM, A.T. 1976. Politics vs. science in food additives regulation. Food Tech. *30,* (8) 56,58,60,62.

THURSTON, R.N. 1977. What consumers should know about foods. Food. Prod. Dev. *11,* (1) 46,48.

3

Fair Packaging and Labeling Act (FPLA)

According to legend, the FPLA was first conceived by Sen. Philip A. Hart (*D*—Mich.) when he became curious about a frozen cherry pie. He wanted to find out whether the pie actually contained the amount of cherries noted on the label of the carton. To his utter amazement, he discovered that the amount of cherries was incorrect. Because of this inconsistency and the sense of a good political issue, he fought for the enactment of the FPLA from 1961 to its eventual passage in 1967. But the climate for legislative control of both advertising copy and deceptive practices was already ripe for a change.

Up until the late 1950's, packaging legislation dealt almost solely with chemicals that may migrate into packaging materials. Labeling legislation was basically concerned with protecting the health of the consumer as well as with preventing deceptive practices. But in the post World War II years, industry had generated additional packaging problems. Deceptive labeling and misinformation were common on the marketplace. Terms such as "Super Family Economy Size" were used for varying actual sizes. Illustrations appeared on products even though they did not reflect the contents of the package. And when Sen. Hart began his investigation of packaging and labeling practices in 1961, he had the full support of consumer groups. The latter felt that this was a sort of "proving ground" as to their strength in public issues.

After six years in Congress, the FPLA was passed in 1966 in an effort to provide consumers with enough information to facilitate "value comparisons" (especially price comparisons) in shopping for consumer goods.

The law became effective in various stages up to 1969. It was specifically designed to protect the pocketbook of the consumer.

FAIR PACKAGING AND LABELING ACT ADMINISTRATION

Administration of the law is divided between three separate government agencies. The FDA controls food, drugs and cosmetics while the FTC has specific control over other "consumer commodities" outside of FDA's jurisdiction. The Department of Commerce (DOC) is responsible for the voluntary standardization of packaging as a means of reducing the sizes currently on the market.

All three agencies are given authority to regulate label information on grocery items and the FDA and FTC are also given permission to regulate non-functional slack fill.

The responsibility for compliance with the law rests with the product manufacturer. Excluded are retailers and wholesale distributors (except where they are product packagers). Exempt from FPLA provisions are meat and poultry products (covered under Treasury Department Laws), prescription drugs, seeds (covered under provisions of the Seed Act) and narcotics and biologicals (specific laws apply).

The *mandatory* provisions of the FPLA include:

(1) Label listing of product identification and form in appropriate type, size and location
(2) Listing of serving sizes and ingredients
(3) Address on label of manufacturer or distributor

The *discretionary* provisions include:

(1) Requirement on package size characteristics
(2) Deals and promotions
(3) Non-functional slack fill control

A violation of the FPLA is considered to be a violation of the specific agency involved. Criminal provisions do not apply. An example would be a complaint issued by the FTC. After a hearing (see Chapter 2) is held, a "cease and desist" order is rendered against the manufacturer. If, after appropriate court review, the order is upheld, then the manufacturer must desist from marketing his product. If he refuses to follow the court order, then he is subject to a fine. Both the FDA and DOC follow similar procedures.

It is this slow procedure that has caused many consumer groups to raise serious objections as to the current FPLA. By the time the FTC or FDA obtains a judgement, the product has already been sold for many months. To the consumer movement, the FPLA is a "law without teeth."

The Magnuson Bill, recently reintroduced into Congress promises to strengthen the weaker provisions of the 1966 FPLA.

FOOD AND DRUG ADMINISTRATION

One of the initial acts carried out by the FDA was to set up regulations that specified the various products under their jurisdiction. A new term was also introduced, i.e., Principal Display Panel (PDP). It is defined as the part of a label most likely to be displayed, shown or examined at the retail sale. The PDP must be able to accomodate all mandatory copy on each panel. Packages with alternate PDP's, or two primary faces, must have duplicate mandatory copy on each panel. On a *rectangular* package, the PDP is considered to be one entire side. Bags forming sides by virtue of the gussets would be considered rectangular and the width of the face times the height of the product is considered the face area. The PDP of a *cylindrical* package is considered to be 40% of the circumference times the height of the container. Excluded from these calcula-

Courtesy of Milprint

FIG. 3.1. PRINCIPAL DISPLAY PANEL

tions are flanges, shoulders and necks of jars and bottles. On *flat* packages, the PDP consists of one entire side. On *all other* shaped packages, the PDP is determined to be 40% of the total surface of the package. This includes package shapes such as spheres and cones.

The identity of the commodity must appear prominently in bold type on the PDP with other printed matter and generally placed parallel to the base of the package. The identity statement must consist of either the product name, common or usual name, or an appropriately descriptive term used by the public for a food when the nature of the food is obvious. The name and place of business may be placed on the PDP or on an adjacent panel. When the name is different from the manufacturer of the contents, a qualifying statement must be used, such as "manufactured for," "packed by," or "distributed by." The statement of the place of business shall include the city, state, and zip code. The street address must be added if a place of business is not listed in the telephone directory. Also on the PDP must appear a net quantity of contents in terms of weight, measure, or numerical count. Fractions or decimals may be used in their lowest form. The declaration shall be in legible type as a distinct item and in contrast to other matter on the package. This declaration must appear within the bottom 30% of the panel, generally parallel to the base. The 30% requirement does not apply to packages having 5 sq in. or less of PDP. It shall be separated from all other printed information above and below by a space equal to the height of the letters used and a space twice the width of the largest letter used on the declaration. No qualifying statements may be used such as giant, jumbo, extra large, etc.

Letters used in the statement must be no more than three times as high as they are wide. Ratio of letter height to size of package is as follows:

(A) Not less than 1/16 in. in height on packages with a principal display panel of 5 sq in. or less
(B) Not less than 1/8 in. in height on packages with a principal display panel of more than 5 but not more than 25 sq in.
(C) Not less than 3/16 in. in height on packages with a principal display panel of more than 25 but not more than 100 sq in.
(D) Not less than 1/4 in. in height on packages with a principal display panel of more than 100 sq in.
(E) Not less than 1/2 in. in height on packages with a principal display pancl of more than 400 sq in.

The letter height pertains to upper case. If upper and lower case is used, the lower case "o" shall meet the standard height. If a fraction is used, each component numeral shall meet one-half of the minimum height standard.

Courtesy of Milprint

FIG. 3.2. STATEMENT OF IDENTITY

Courtesy of Milprint

FIG. 3.3. DUAL EXPRESSION

On packages containing less than 4 lb or one gal., the declaration shall be expressed in both ounces followed by pounds with the remainder in ounces. In a package of 4 lb or one gal., the expression of only ounces may be eliminated. Proper abbreviations may be used and copy can be on more than one line.

Examples: (A) 1½ pounds weight may be expressed as "Net Wt 24 oz (1 lb 8 oz)" or "Net Wt 24 oz (1.5 lbs)" ¾ pound avoirdupois weight shall be expressed as "Net Wt 12 oz."

1¾ quarts may be expressed as "Net Contents 56 fluid ounces (1 quart, 1½ pints)" or as "Net 56 fluid oz (1 qt 1 pt 8 oz)."

Metric units are optional. Additional information may be put on a supplemental panel called the *informational panel.* If the top panel is the PDP, then any adjacent panel can be considered to be the informational panel.

This concept went into effect in 1973; however, the final deadline for compliance was as late as 1975. There are certain exceptions to these definitions:

(1) *Milk, water and soft drinks* — information is usually found on cap or sloping shoulder of milk containers
(2) *Multipacks*
(3) *Retail wraps*
(4) *Basket carriers for soft drinks*
(5) *Small cosmetic packages (below 5 sq in.)* — information must be on card to which package is fastened
(6) *Individual serving* — size packages of foods containing less than ½ oz or ½ fl oz for institutions or restaurants and not intended for retail sale
(7) *Individually wrapped pieces of candy* — when such candy is in an outer container or bag carrying the mandatory copy

No qualifying statements or words may be placed in an area where mandatory copy is used. If there are statements, such as "family size," on the PDP, they can be regulated by the FDA if and when these terms get out of control.

A number of servings statement may be found on the informational panel. This must be in the same size type as a statement of the net quantity of each serving. This can be expressed in different terms than the requirements for net quantity of a package. The generic form of the product may be on the informational panel. It must be in prominent type and be as bold as the trade name.

The requirement on ingredient labeling caused a significant amount of initial confusion to FDA personnel. Noted in the FPLA is the statement that each label must contain a list of ingredients in decreasing predominance on any appropriate informational panel in adequate type.

The entire ingredients statement shall appear on a single panel of the label. Conbinations of two like ingredients such as cottonseed oil and olive oil must designate the percentage of each. This is valid unless the product is covered by a Standard of Identity Regulation. To date, the deviations to the rule of decreasing predominance apply only to products containing expensive ingredients, such as maple flavored syrup. Because of the fairly expensive cost of pure maple syrup, it is necessary to state the percentage of maple syrup contained in the product on its label. Preservatives and their function must also be listed. Listing of fats and fatty acids are optional; however, when listed they must follow a set of specifications issued to standardize this type of labeling. Special dietary statements are also found on the informational panel.

Illustrations appearing on packages have also been codified. Any main dish on a package must boldly state any added ingredients necessary to make the dish illustrated. Examples are various types of food extenders.

On dual statements, wheat flour in sizes of two lb or more are exempt from a dual statement. Milk, milk products, fruit juices, water, and soft drinks also are exempt from dual statements. Ice cream is exempt in the standard size. If it is in a package containing a fractional size, a dual statement of size is mandatory. Egg cartons are exempt from the 30% requirement.

Courtesy of Milprint

FIG. 3.4. EXEMPTIONS

In the cosmetic area, it is required that fractional quantities in metrics be listed as decimals to two places. Also, for packages less than ½ oz, the provisions of the FPLA do not apply if the small package is attached to a card which contains all mandatory information.

The 1938 Act required the listing of active and certain specified ingredients for drugs. The FPLA authorized the FDA to require the listing of all inactive ingredients for drugs. The FDA has not proposed that this be done. It might be useful to require a statement of the general pharmacological action of each active ingredient in over-the-counter drugs as the names are not meaningful to most consumers. Prescription drugs only require a statement of each active ingredient in the drug. Exempt drugs include sample packs of ointments less than 8 gm and liquid nonprescription veterinary products (no dual statement required).

FEDERAL TRADE COMMISSION

Regulations administered by the FTC are essentially similar to those of the FDA; however, because of the type of products covered under FTC regulations, a bidimensional requirement is necessary. (See Table 3.1 for governmental organization.) Sheet type products are expressed both in sq ft and sq in. For consumer bags, it is required to list the number of bags in the package as well as their dimensions. This is also true for disposable paper products. There is an entire list of exemptions:

(1) Photographic film
(2) Christmas tree ornaments
(3) Vacuum cleaner bags
(4) Sponges
(5) Paper tablecloths
(6) Chamois cloths
(7) Candles
(8) Solder

Promotions are limited to three per year by FTC regulations and it must be a real offer or an offer of "tangible value."

When manufacturers label products "X Cents Off" and give a discount to retailers, they employ a promotion device that has been effective in stimulating sales, particularly of new products. This promotion device has been misused by some manufacturers who make the offer so frequently, or for such an extended period of time that the "reduced" price becomes the regular price. Hence, the offer is not a genuine price reduction. In an effort to curb such abuses, the FTC in 1967 investigated "cents off" promotions by instant coffee producers and threatened to initiate proceedings against firms who abused the promotion device. The

TABLE 3.1. U.S. GOVERNMENT ORGANIZATION

(A) The Legislative Branch
- (1) The Congress
  - (a) the Senate
  - (b) the House of Representatives
- (2) Congressional committees relating to foods
  - (a) of the Senate
  - (b) of the House of Representatives

(B) The Judicial Branch
- (1) The Supreme Court
- (2) The U.S. Appellate Courts
- (3) The Federal District Courts

(C) The Executive Branch
- (1) The President — The "Chief Executive"
- (2) The Cabinet — The Cabinet Departments
- (3) Other independent agencies
- (4) Quasi-government agencies

Department of Agriculture requires a showing that the "cents off" is going to the consumer and that a statement of the regular price be made before it approves a "cents off" label.

DEPARTMENT OF COMMERCE

The DOC never really desired the task of deciding exactly how many package sizes consisted of "undue proliferation"; however, the FPLA provides that the Secretary of Commerce, rather than the regulatory agency, make the determination that there is undue proliferation of sizes for a particular commodity.

Any products, currently in the marketplace, are sold in a wide variety of sizes, shapes and forms. The DOC attempted to use their mandate in two specific commodities, i.e., toothpaste and potato chips. They did succeed in removing from sale the sample size tubes of toothpaste. But 10-12 different size tubes are still on the market. Potato chip packages were carefully examined and several sizes were removed from the market. Other sizes remained and later, new sizes were introduced. Over 71 different sizes of packages for potato chips are now for sale. These were the only two cases tried by the DOC and test products remain in the DOC's responsibility area. Because of the almost total lack of enforcement by the DOC as to size proliferation, consumers are now agitating for a revision of the FPLA (the Magnuson Bill now in Congress). Labeling hearings will probably he held immediately followed by hearings on a wide variety of subjects.

It is interesting to note that in Sweden, sizes have been standardized and regulated. Descriptive labeling has been fully standardized in other European countries. In the United States, these concepts are still in their infancy.

NUTRITIONAL LABELING

Guidelines have already been formalized for nutritional labeling but were deferred because of the poor state of the U.S. economy.

Initiated as a voluntary program by the FDA in 1969 because of results obtained from the White House Conference on Food and Nutrition, nutritional labeling was made mandatory and regulations published in 1972. Compliance was postponed in 1975 and again in 1976. As published, the regulations require nutritional data on all foods. An interesting provision states that if an established imitation food has been proven to be equal or superior to a natural food, the term "imitation" may be omitted from the label. Nutrients are listed as the percentage of the recommended daily allowance (RDA) and they are categorized as macronutrients and micronutrients. Macronutrients include calories, proteins, fats and carbohydrates. Micronutrients include vitamins and minerals.

REFERENCES

ANON. 1975A. Washington survival guide. Pkg. Eng. *20* (5), 36-39.

ANON. 1975B. A guideline to interpretation of packaging and labeling regulations. Can. Pkg. *28*, (10), 63.

ANON. 1975C. Need an hour's heavy reading? Proposed food labels could serve. Food & Drug Pkg. *32* (7), 17, 15.

ANON. 1976A. Fair packaging and labeling: A capsule report by Milprint, Inc. The Mfg. Confec. *56* (9), 24-28.

ANON. 1976B. Packaging abroad. Packaging and labeling requirements. Containers & Pkg. *28*, (4), 15.

ANON. 1976C. FDA recall guidelines require codes on all food, drug, cosmetic products. Food & Drug Pkg. *35*, (3), 31.

BAILY, R.F. 1975. The good and bad of food labeling. Pkg. Dig. *12*, (9), 15, 18-19.

BAUMAN, H.E. 1976. Cost impact of food regulations. Food Tech. *30*, (8), 50, 52, 54.

HOPKINS, H. 1976. Getting specific about fats and oils. FDA Consumer *10*, (2), 13-14.

LABUZA, T.P. 1976. Regulation of nutrition or nutrition of regulation. Canner & Packer *145*, (6), 46, 57.

LEHMANN, P. 1976. Getting the word from consumers. FDA Consumer *10*, (3), 9-13.

MILLER, F. 1975. GAO recommends Encyclopedic Food Labels., *32*, (7), 1, 17.

SACHAROW, S. 1976. A guide to government packaging regulations. Part 2. Labeling regulations. Pkg. Printing & Diecutting *22*, (3), 8, 10, 77.

SPEER, J.F. 1975. Package labeling. Am. Dairy Rev. *37*, (7), 43-45.

STEVAN, M.A. 1974. What shoppers are concerned about. FDA Consumer *8*, (7), 13-17.

STOUFFER, L. 1976. A look back into the future of packaging regulations. Pkg. Digest *13*, (7), 80-81.

4

United States Department of Agriculture

Abraham Lincoln signed the congressional act creating the U.S. Department of Agriculture (USDA) in 1862. A little more than a quarter century later, agriculture became the eighth executive department to achieve cabinet rank. In those early days, the department served a population almost entirely engaged in agriculture as a livelihood. A farmer in Lincoln's time produced only enough farm products to supply himself and three or four others.

Today's department serves a nation in which less than 5% of the population lives on farms — a nation in which each agricultural worker provides food and fiber for himself and over 50 others.

In 1862 Congress passed, and President Lincoln signed, the Land-Grant College Act. This act donated 11 million acres of public lands to the states and territories to provide colleges for the benefit of agriculture and the mechanical arts.

In 1887 Congress authorized an agricultural experiment station in each state and territory. In 1914 it established the Cooperative Extension Service for each state to extend agricultural and home economics research information of the department and land-grant colleges to farmers and other people.

These acts were followed by others to strengthen the college teaching, experimental, and extension activities of the states.

These agencies, in cooperation with the department have made American agriculture the most productive, and the consumer's food the least expensive in terms of disposable income, the world has seen. The land-grant colleges and universities, totaling 70, with at least one in every State, Puerto Rico, and the District of Columbia, in cooperation with the Extension Service and experiment stations, work to improve every aspect of agricultural production and distribution.

ACTIVITIES OF USDA

USDA works closely on research with state experiment stations, state departments of agriculture, schools of forestry, farmer cooperatives, and other public and private agencies. It administers Federal grant funds voted by Congress for the states. (See Table 4.1 for USDA organization.)

TABLE 4.1. USDA ORGANIZATION

SECRETARY
Under Secretary
Deputy Under Secretary
Staff Assistants

- Judicial Officer
- Office of Communication
- Office of the General Counsel
- Office of Investigation
- Office of Intergovernmental Affairs
- Office of Planning and Evaluation
- Departmental Administration
  - Assistant Secretary for Administration
    - Office of Administrative Law Judges
    - Office of Audit
    - Office of Management and Finance
    - Office of Equal Opportunity
    - Office of Automated Data Systems
    - Office of Operations
- Agricultural Economics
  - Director
    - Staff Economist
    - Farmer Cooperative Service
    - Statistical Reporting Service
    - Economic Research Service
- International Affairs and Commodity Programs
  - Assistant Secretary
    - Agricultural Stabilization and Conservation Service
      - Commodity Credit Corporation
    - Federal Crop Insurance Corporation
    - Foreign Agricultural Service

TABLE 4.1. *(CONTINUED)*

Marketing and Consumer Services
  Assistant Secretary
    Agricultural Marketing Service
    Animal and Plant Health Inspection Service
    Commodity Exchange Authority
    Food and Nutrition Service
    Packers and Stockyards
      Administration
Rural Development
  Assistant Secretary
    Farmers Home Administration
    Rural Development Service
    Rural Electrification Administration
      Rural Telephone Bank
Conservation, Research and Education
  Assistant Secretary
    Agricultural Research Service
    Cooperative State Research Service
    Extension Service
    Forest Service
    National Agricultural Library
    Soil Conservation Service

The Agricultural Research Service carries out research on crops, livestock, soil and water conservation, agricultural engineering, control of insects and other pests, human nutrition, and home economics. ARS also develops new and expanded uses for farm commodities; it conducts and administers a research program using the physical and biological sciences to solve problems of market quality, transportation, and facilities.

The Cooperative State Research Service administers (under the Hatch Act of 1955 and the McLintire-Stennis Cooperative Forestry Research Act of 1962) federal grant payments for research at the agricultural experiment stations and eligible schools of forestry. CSRS also assists state experiment stations, land-grant colleges, and USDA agencies in planning and coordinating scientific research programs.

The Economic Research Service does research in general economic and statistical analysis, marketing economics, farm and resource economics, and foreign economic analysis.

The Farmer Cooperative Service conducts economic research to help farmers market their products and purchase supplies. It also helps all rural people to obtain other business services through cooperatives.

The Farmer Cooperative Service reaches farmers and other rural people by conducting educational work to help them improve the effectiveness of their cooperatives. In these activities it cooperates with land-grant colleges, state extension offices, county agents, and cooperatives and their state and national organizations.

The Forest Service carries on a broad research program through its eight regional forest experiment stations, the Forest Products Laboratory, and the Institute of Tropical Forestry. In these laboratories, scientists study the establishment, improvement, growth, and harvesting of timber; protection of forests from fire, insects, diseases, and animal pests; management of rangelands; improvement and management of wildlife habitat; forest recreation; protection and management of watersheds; efficient marketing and utilization of forest products; forest engineering; forest genetics; and forest economics. A continuing forest survey provides comprehensive information on the extent and condition of forest lands, the volume and quality of timber resources, trends in timber growth and harvest, and the outlook for future supplies and demands.

Courtesy of U.S. Dep. Agric.

FIG. 4.1. USDA INSPECTOR
USDA inspector checking equipment for sanitation levels.

Education

The Cooperative Extension Service is the field educational arm of the Department of Agriculture and land-grant universities and colleges. Federal, state, and local governments jointly sponsor and finance its work. The Extension Service staff represents USDA in this activity.

In nearly all countries, extension workers conduct educational programs to help families learn about and apply new research findings and technological developments to the everyday problems of living and making a living. The major areas included in the educational efforts are efficient production and marketing of agricultural products; improved family living, including nutrition; 4-H youth development; and rural development.

The National Agricultural Library, Beltsville, Maryland (intersection of I-495 and U.S. 1 and Exit 27, North) contains 1.3 million volumes on agriculture, biology, chemistry, and other subjects supporting agricultural research. Probably the most extensive agricultural library in existence, it collects and maintains worldwide publications in the agricultural, biological, and chemical sciences; serves the research, extension, regulatory, and other programs of the USDA and state agricultural agencies; extends services to other libraries, institutions and the public; and provides, at cost, microfilm and photocopy of material in the collection.

Marketing and Consumer Services

The Commodity Exchange Authority keeps watch over and investigates traders' and brokers' operations on commodity exchanges to prevent price manipulation and fraud and to safeguard producer and consumer interests in pricing and marketing services of exchanges.

The Agricultural Marketing Service (AMS) helps the private marketing system move food and other farm products from producer to consumer quickly and efficiently, with fairness to all, through a group of related programs.

AMS establishes standards for grades indicating the quality of a large number of farm products. It provides grading services for many products — most of them voluntary and paid for by the person or firm requesting the service. Similarly, it operates acceptance services to make sure that commodities meet the contract specifications agreed to by individual firms. And it collects market news daily on farm product prices and movement in the major markets for various qualities of products.

AMS strengthens farm income and expands the market for farm products by promoting increased consumption of foods in temporary overabundance, and by purchasing foods in excess supply for distribution through programs of the Food and Nutrition Service.

AMS helps farmers get reasonable rates and services for transportation of their products by presenting information to the Interstate Commerce Commission. It also promotes fair play in marketing through administration of the Perishable Agricultural Commodities Act, the U.S. Warehouse Act, the Federal Seed Act, the Plant Variety Protection Act, and the Agricultural Fair Practices Act.

Through marketing agreements and orders, AMS puts a floor under prices received by farmers for milk, and helps regulate the volume or quality of fruits and vegetables coming to market from areas and commodities coming under the orders.

AMS also helps state departments of agriculture develop better ways of marketing through the Federal-State Marketing Improvement Program. In fact, many AMS programs are carried out in cooperation with the states. The agency also published marketing guides for several commodities to help producers bring supplies into balance with demand at a reasonable price.

Consumers also benefit from many of these programs that help to assure adequate, dependable supplies of food and other farm products, marketed quickly and economically.

The Food and Nutrition Service leads the nation's effort to wipe out poverty-related hunger and malnutrition. Its Food Stamp and Food Distribution Programs help provide food to poor families, both through the regular food marketing system and through a direct distribution system. Its Child Nutrition Programs help to feed children through the National School Lunch Program, School Breakfasts, Special Milk, and other child-feeding activities in preschool centers and summer programs. When hurricanes and other natural disasters strike, resources of all these FNS programs are quickly adapted to aid its victims.

The Animal and Plant Health Inspection Service (APHIS) helps to safeguard the health and quality of the nation's animals and plants, and protect consumers by maintaining a strong system of Federal-State inspection of meat and poultry products for wholesomeness and truthful labeling.

Protection of our farm and food products begins at the ports of entry, where APHIS maintains close surveillance of all animal and plant imports, cargoes, and passengers' baggage to prevent the entry of anything that might harbor a foreign pest or disease of animals and plants.

APHIS cooperates with the states and the agricultural industry to eradicate and control animal diseases of national importance. In addition to the protection this provides to the nation's farmers, it helps to guard all Americans from those diseases which can be transmitted from animals to man.

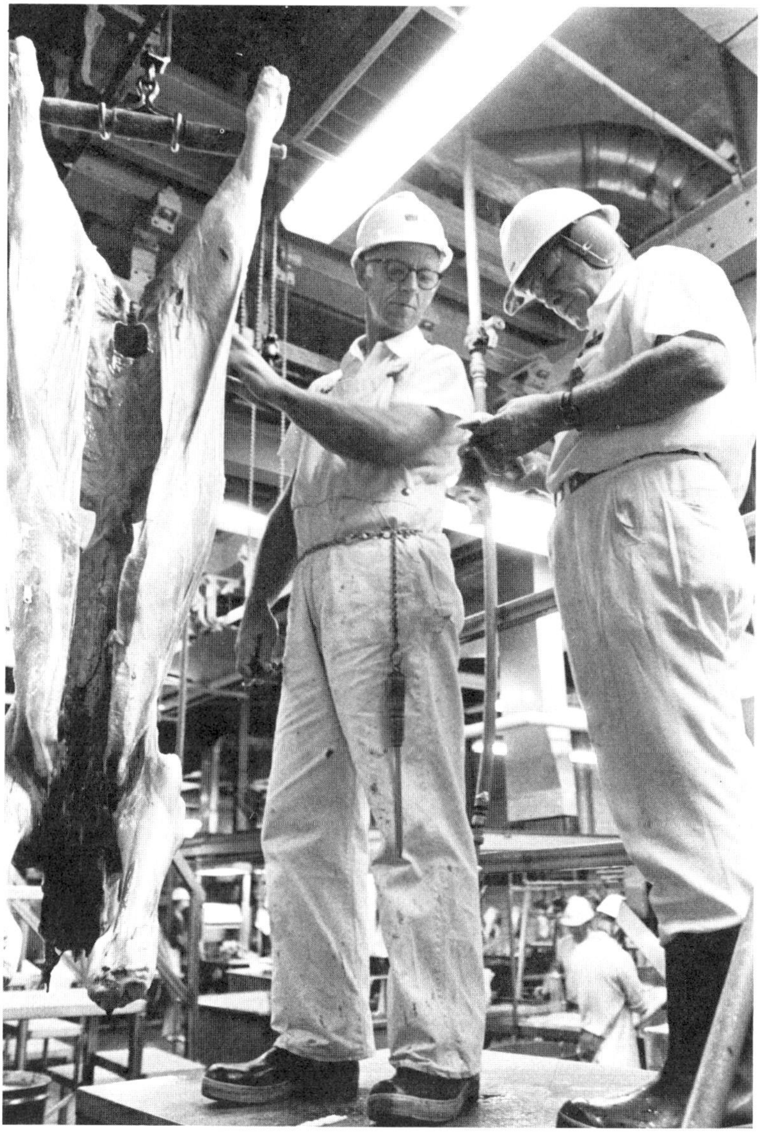

Courtesy of U.S. Dep. Agric.

FIG. 4.2. MEAT INSPECTION
USDA inspector checking quality of domestic meat.

The agency also cooperates with the states and industry to control or eradicate native or alien pests and diseases of plants which pose an interstate menace. Close surveillance is maintained nationwide of pest infestations to identify potential problems and organize effective control programs.

Courtesy of U.S. Dep. Agric.

FIG. 4.3. POULTRY LINE
Poultry products are checked by USDA inspectors to insure quality.

Typical of these cooperative activities are the current campaigns to control the gypsy moth, which is capable of defoliating shade trees and forests; the imported fire ant, a severe nuisance to animals and humans; the screwworm, a flesh-eating menace to livestock, wildlife, and all other warm-blooded animals; and the cereal leaf beetle, which can cause severe damage to small grains.

APHIS also licenses the manufacture and marketing of veterinary biologics to ensure their purity and effectiveness. In its role as a regulatory agency, it enforces the laws requiring humane treatment of dogs, cats, and other small animals used for research, zoo animals and those used in exhibitions, and horses.

The agency further administers the Federal Meat Inspection Act and the Poultry Products Inspection Act, which, together, provide for a uniform federal-state system of inspection for all meat and poultry products. Federal inspection requires advance approval of the construction and equipment of all plants operating in interstate commerce; continuous inspection of all processing operations; thorough inspection of all animals and birds before, during and after slaughter; and advance approval of all packages and labeling of meat and poultry products. It also includes surveillance to guard against illegal residues of chemicals and inspection of all imported meat and poultry comparable to that given to domestic production.

In addition, the laws require that states must maintain inspection systems for meat and poultry produced and sold within their own boundaries that are equal to federal inspection, or USDA must assume responsibility for all inspection in a state.

The Packers and Stockyards Administration administers the Packers and Stockyards Act, helps to maintain free and open competition in the marketing of the nation's livestock, poultry, and meat production.

The P&S Act prohibits unfair, deceptive, discriminatory, and monopolistic marketing practices which would deprive the farmer of a fair price for his product, deprive the marketer of a reasonable profit for his services, and would unduly affect consumer prices for meat and poultry.

PACKAGING[1]

The safety of materials for packaging and handling of meat and poultry food products is part of the responsibility assigned to the Meat

[1]From "Materials Evaluation and Acceptance" USDA, APHIS.

and Poultry Inspection Program by the Federal Meat Inspection Act as amended by the Wholesome Meat Act of 1967 and by the Poultry Products Inspection Act as amended by the Wholesome Poultry Products Act of 1968. The safety is assured by examination of materials composed of substances which could migrate to the food by diffusion, dissolution, absorption, accidental abrasion, or other means of transfer. The materials may become part of a container, a piece of equipment, a structure, or a process, each of which may require physical or other evaluation in addition to this chemical evaluation. Thus, the Chemistry Staff examines materials for substances which are toxic or hazardous or which could migrate in undesirable amounts under the actual conditions of use. This finding is then reported, free of charge, to the supplier and other appropriate program functions. Acceptable cleaning and sanitizing agents will be shown in the list of Chemical Compounds and other acceptable materials will be named in a letter.

Container materials findings are reported to Labels and Packaging Staff for final action on the container. Likewise, equipment materials findings go to the Equipment Group, structural materials findings to the Facilities Group, branding inks findings to the Meat Group for final action on acceptability of the finished item.

Requirements

Three pieces of information must be supplied for evaluation of a material or group of materials:

(1) Brand Name or manufacturer's identifying mark;

(2) Description of the conditions of use;

(3) List of substances composing the material.

The list must identify all major and minor constituents by proper chemical name as it appears in the appropriate Food Additives Regulation, or by reference to previous acceptance. Trade secrets and proprietary information are held in confidence and used only to evaluate the material. The information should be sent to:

Compound and Packaging Group
Chemistry Staff
Scientific Services, APHIS
U.S. Department of Agriculture
Washington, D.C. 20250
Telephone: (202) 447-7680

Immediate Containers for meat or poultry food product must be fabricated of only those materials in compliance with the Food Additives Regulations, particularly 21 CFR 121 Subparts B (Exemptions), E (Prior

Sanctions), and F (Containers). It is important to note the type of food in direct contact (aqueous and fatty) as well as the temperatures to which the packaged food will be exposed. "Below 150°F" includes frozen storage, room temperature storage and shrink tunnel treatment, where the temperature of the food product does not rise appreciably. "Below 250°F" includes these plus boil-in-the-bag, heat sterilization, retorting, etc. "Oven temperatures" includes all these plus roasting or cooking above 250°F. Plastic resin films must comply with the appropriate Food Additives Regulation extractability requirements. Adhesives for laminating and fabricating must comply with 21 CFR 121.2520 and pressure-sensitive adhesives for band labels, giblet wraps, etc., must comply with 21 CFR 121.2577. Complete, whole containers must be cleared by the Labels and Packaging Staff.

Inks and colorants are evaluated according to their use. Branding Ink, for direct and prolonged contact with meat or poultry, is composed of water, alcohols, sugar, thickeners, solvents, and other GRAS Substances with FDA certified food colors. Official Branding Inks must also be cleared by the Meat Group, Inspection Standards and Regulations Staff.

Insert Label Inks, also for direct and prolonged contact with meat and poultry, must comply with 21 CFR 121.2514 plus FDA certified food colors whether coated with overprint varnish or not.

Package and label inks (and overprint lacquer) for the outside of immediate containers must have a vehicle in compliance with 21 CFR 121.2514 and pigments chosen from that Regulation or from FDA Provisional Lists of Food Colors or from dyes and pigments accepted by Scientific Services. The following are specifically excluded: Lakes or pigments containing Antimony, Arsenic, Cadmium, Chromium, Lead, Mercury and other toxic metals; flourescent brighteners; known carcinogens, mutagens, and teratogens. Colorants for plastic films, tubs, containers, etc., are subject to the same restrictions. Dyes and pigment should always be identified by their 5-digit Colour Index number or structural formula.

Equipment Materials are evaluated on the basis of their chemical composition. Small equipment, such as aprons, bags, cutting boards, gloves, pins, skewers, timers, tools, twine, etc., is handled the same as immediate containers. Larger equipment, such as conveyor belting, hose, metals, metal coatings, pipe, plastics, rubbers, solders, etc., must have its chemical composition cleared and then be referred to Equipment Staff, Plant Facilities and Equipment Staff, Technical Services Staffs, APHIS. Paints and coatings applied to equipment, machinery or panels before installation are evaluated similarly to inks by this office.

Facilities Materials such as caulking, insulation, panelling, structural members, etc., are evaluated chemically by this office, but must then be referred to Facilities Staff, Plant Facilities and Equipment Staff, Technical Services Staffs, APHIS.

Cleaners and Sanitizers, boiler water treatment, pesticides, lubricants, etc., and those paints and coatings to be applied *on site* are evaluated by the Compounds Evaluation Laboratory. All except paints, when found acceptable, will be listed in the List of Chemical Compounds, Request Forms and instructions from:

Compounds and Evaluation, Chemistry Staff
Scientific Services, APHIS
U.S. Department of Agriculture
Building 306, ARC
Beltsville, Maryland 20705
Telephone: (301) 344-2566

Courtesy of U.S. Dep. Agric., Photo by Jack Schneider

FIG. 4.4. MEAT INSPECTION
A new Federal meat inspector (right) works his first few months at a meat plant under the watchful eye of an experienced instructor.

General Considerations.—The APHIS has no authority to evaluate materials for use with foods other than federally inspected meat and poultry. Materials which meet the APHIS criteria are termed "Acceptable" because the term "approved" might erroneously indicate a preference for one brand over another. The phrase "accepted by APHIS,

USDA" may be used in brochures, catalogs, and letters to customers where valid, but is inappropriate for use in advertising.

Manufacturers requesting acceptance of many similar items such as a system of inks or paints differing only in pigment, gloves of the same material but different style, cartons of the same composition but various sizes, film of the same resin blend but different gauge, should group them in a convenient way to facilitate the acceptance. Materials not specifically described here or in the Food Additives Regulations must be considered individually and regulated in a way most likely to guarantee the consumer complete protection against toxic or hazardous substances. If a material is found to contain an unacceptable substance, the petitioner will be so advised and allowed to make a substitution or withdraw the request. If an acceptable substance is substituted in a formulation previously found acceptable, notify the APHIS so that the old letter of acceptance can be validated or a new one issued. Once an evaluation has been made and materials found acceptable, one final criterion must be met: The material, when placed in actual use, must perform in a safe and satisfactory manner.

Labels for USDA Approval

Since labels for meat and poultry products must be approved by USDA personnel, the method used is of major importance.

The following briefly outlines a pattern for expediting the review of labels. The directions are designed to enable an official establishment to obtain an approved sketch and/or final label in the shortest time possible.

All sketches and labels submitted for an approval should be mailed to:

Benjamin Franklin Station
P.O. Box 7416
Washington, D.C. 20044

No additional information should be included with the above address. An "Attention Mr. X" line only slows the distribution of the incoming labels. If a label is to be submitted to a specific person, the person's name should appear on a paper attached directly to the transmittal form inside the mailing envelope. *Special delivery* delays the delivery of labels.

Procedures for Submitting a Sketch.—The sketch should be submitted in triplicate. Each of the three copies is to be attached to a transmittal form. The three copies may be typewritten, prepared as a printer's rough, paper takeoffs, or other legible forms.

Show all required labeling features including the following:

(1) Product name;
(2) Ingredient statement and/or qualifying statement;
(3) Inspection legend;
(4) Firm name and address including zip code;
(5) Refrigerated or frozen statement (if necessary);
(6) Net weight statement for products sold at retail intact.

Show exactly how label is to be used, including type and size wrapper, casing, carton, etc. State whether it is a shipping container or consumer size container to be sold at retail intact. Include size of principal display panel.

With each sketch show complete formula including a brief method of preparation. Any additional explanation of formula, such as listings of spice and flavoring mixtures, may go in "remarks" box.

The transmittal form should be signed by the plant representative and submitted through the inspector in order to benefit from his comments and recommendations. The sketch should then be ready for mailing. Sketch approvals should be received from the Washington office at the official establishment before the final label is ordered from the printer.

Procedures for Submitting a Label.—Submit three copies of actual label, each attached to transmittal form. Submit four copies of Foreign Labels plus one copy for each port of entry. Duplicate *all information* as recommended for sketch approval. Also include the date of the sketch approval from Washington.

It is advisable to submit samples of products for which we have no published or unpublished standard in order that they may be reviewed. Such products should be mailed to the following address:

Product Standards Staff
Technical Services, APHIS
U.S. Department of Agriculture
Washington, D.C. 20250

Please arrange a mailing schedule that will allow products to arrive during the work week. Very specific printer's instructions are necessary, particularly when an official establishment is ordering an approved sketch to be printed as a final label. When submitting a previously approved label for a new approval, include the original approval number for the label.

The following are common errors or deficiencies that result in the disapproval of a label or temporarily approving a label:

(1) Product name not complete or acceptable;
(2) Product name not shown with required prominence;

(3) One or more required labeling features not shown with acceptable prominence;
(4) All required labeling not shown on principal display panel;
(5) Firm name and address not shown as authorized for each establishment;
(6) If packed for another firm, failure to precede the firm name and address with the term "Packed For" or "Distributed By";
(7) Ingredients statement does not agree with formula shown on transmittal form;
(8) Ingredients not listed by their true name;
(9) List of components not preceded by the word "Ingredients" followed by a colon;
(10) Incomplete formulas and methods of preparation;
(11) Not showing type of container, casing, etc.;
(12) Not showing if label is for consumer size, institutional or shipping container;
(13) A good "rule of thumb" for inspection legends:
  (a) Show "U.S." and "Est." twice the height of intervening logo;
  (b) Show intervening logo 3/5 the width of "U.S." and "Est.";
(14) Net weight not shown with required prominence;
(15) If catch weight (pounds — ounces) should be shown with net weight or applied mechanically and so stated on transmittal form;
(16) Coloring of labels should not be of such color, design or kind to be misleading in respect to quality or kind or product;
(17) Illustrations not representative of contents of container;
(18) The statement "Serve as Illustrated" should be shown when anything other than product enclosed is shown in illustration;
(19) "Serve as Illustrated" featured with illustration does not relieve responsibility that product illustrated does not represent product enclosed;
(20) In the case of lithographed can labels do not submit the cans themselves; submit paper take offs in color.

FUTURE OF USDA

Overlapping and inequity in federal regulation on packaging industries are likely to be an early target of the Carter Administration. The need for simplification of this bureaucratic jungle is evident when one sees that plants processing meat into canned and frozen products are under the constant supervision of the Department of Agriculture, which has an army of 8736 inspectors and supervisors and an annual budget of $228 million.

By contrast, the Food and Drug Administration, with responsibility for the remaining and much larger portion of food supply, has only 900 inspectors, and they must spread their time among non-meat foods, drugs, cosmetics, and medical devices. FDA's budget for inspection is $66 million.

Some plants packaging both meat and non-meat products are answerable to both agencies.

"Having a single agency inspect all food would seem to make more sense," said Alexander Schmidt, former chief of the FDA. Both agencies are in agreement that their separate food inspections should be ended. Whether a single inspection service would save money or lessen problems is debatable, but that would seem to be a step in the right direction.

Courtesy of U.S. Dep. Agric.

FIG. 4.5. RESIDUE TESTING
Samples are being collected for residue testing.

REFERENCES

ANON. 1974. Packaging in a tougher world. Mod. Pkg. *47,* (12) 8-10.

ANON. 1975A. USDA regulation on jar closures extended to 1977 as industry reacts. Pkg. Eng. *20,* (11) 25.

ANON. 1975B. Washington survival guide. Pkg. Eng., *20,* (5) 36-39.

ANON. 1975C. Packaging of meat and meat products. Packaging *46,* (45) 48-49.

ANON. 1975D. Trends in vegetable packaging in the U.S.A., Aust. Pkg. *23,* (5) 101.

ANON. 1976. Nutritional analysis key to . . . nutritional labeling. Food Proc. *37,* (1) 47-49.

GAST, L.L. 1975. We need your ideas on labeling policies. The Nat'l Prov. *172,* (1) 94-96.

SACHAROW, S. 1976. Guide to government packaging regulations: Department of Agriculture. Pkg. Prtg. and Diecutting, *22,* (7) 6-7.

SMELING, JR., H.V. 1976. Special report: Food processing's guide to government agencies. Food Proc. *39,* (7) 24-28.

5

Product Safety — Consumer and Industrial

CONSUMER

Regulations dealing with consumer product safety date back to the Middle Ages. In the 15th Century, Scottish authorities passed a law making it subject to treason to take home "any poison harmful to Christian man." In the United States, Congress had always wanted to safeguard the public and particularly children from harmful substances. The *Caustic Poison Act of 1921* was one of the earlier pieces of legislation. But the modern age of product safety regulations began in the early 1950's with the passage of the *Flammable Fabrics Act (FFA) of 1953.* Although this law was nebulous in nature, it did call attention to a significant product problem. Consumer interest soon accelerated and in the late 1950's, the FDA published many articles on varying product safety tests. Reports of children being accidently poisoned by excessive aspirin and lye intake became public knowledge. The *Federal Hazardous Substances Act (FHSA) of 1960* followed and administration of both this regulation and the FFA was given to the already overburdened FDA. Initially, the FFA was to be administered by the FTC.

The FDA acknowledged the problem associated with the accidental poisoning of children and zeroed in on aspirin, detergents, and household chemicals as the three prime products to be regulated. A voluntary commitment was obtained from the firms involved to standardize the strength of childrens aspirins to 1.25 gr and also to limit the quantity sold in one bottle. A child could thus conceivably ingest an entire bottle of children's aspirins and it would not prove to be fatal. Work also began on the development of safety closures for different products. Congress soon felt that stronger legislation was required to supplement the FHSA and in 1966 passed the *Child Protection Act* as an amendment. The *Child Protection and Toy Safety Act* followed in 1969 as an additional amendment to the original FHSA. The latter was prompted by the many problems inherent with the safety of cheap imported toys. The *Refrigerator Safety Act* also was passed as a separate law and it regulated methods of disposing refrigerators without causing child safety problems.

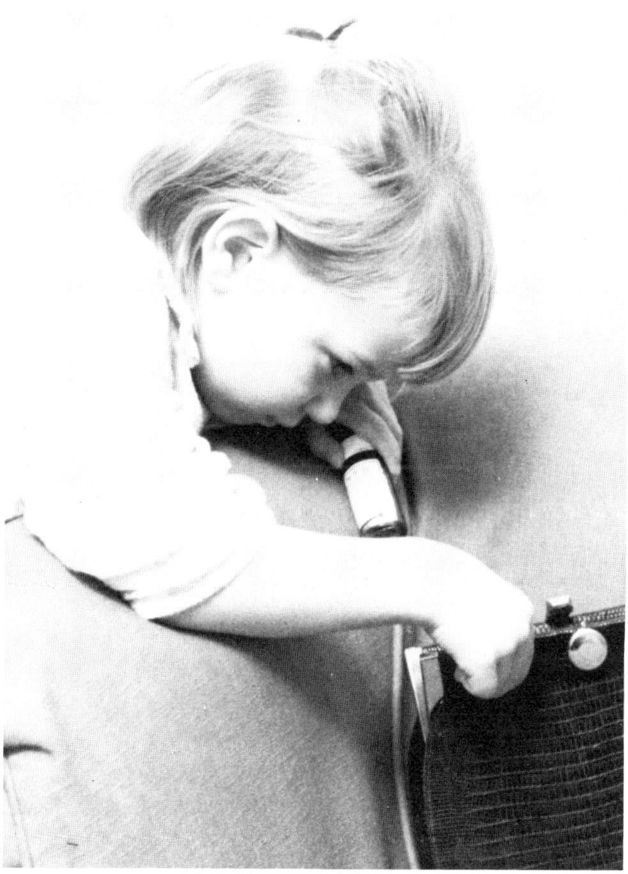

Courtesy of Consumer Product Safety Commission

FIG. 5.1. CHILD IN MOTHER'S HANDBAG
Child is taking medicine vial from mother's handbag. This provoked the need for child-resistant closures.

Courtesy of Consumer Product Safety Commission

FIG. 5.2. LOOK FAMILIAR
An overcrowded closet of old paints, thinners and solvents—a lit cigarette or match could ignite the flammable vapors here and start a fire.

In 1970, the *Poison Prevention Packaging Act (PPPA)* passed and dealt with child resistant closures on numberous products. On October 27, 1972, President Nixon signed into law the *Consumer Product Safety Act.* This Act created a permanent Consumer Product Safety Commission (CPSC), which consisted of five commissioners appointed by the President with the advice and consent of the U.S. Senate. The Commission was designed to protect consumers from unreasonable risk of injury from products. The CPSC is now able to subject products to bans, safety standards, and testing. Also included in the 1972 legislation is the establishment of a 15-member advisory council to assist the Commission. The council includes five representatives from government agencies, five from product industries, and five from among consumer organizations, community groups, or recognized consumer leaders. The functions of the Secretary of HEW under the PPPA are now transferred to the CPSC. All enforcement of the PPPA is now the responsibility of the CPSC. Meetings of the CPSC are to be held no less than four times annually.

Several important characteristics are inherent in the laws mentioned. Except for the PPPA, the above regulations deal only with retail consumer goods and do not include pesticides, foods, drugs, or cosmetics. Products regulated are mainly household chemicals, fabrics and toys. The PPPA extends safety closures and child-resistant packaging to drugs. Not covered in these regulations is the bulk shipment of interstate goods.

Detailed information about the regulations of the CPSC is found in the *Code of Federal Regulations,* No. 16, *Commercial Practices,* parts 150-end. This is now available in a revised edition as of February 1, 1976. Data relative to the PPPA and child-resistance is located in the proceedings of the American Society of Testing Materials (ASTM), Subcommittee 10.32 which deals with safety closures and unit-dose packaging.

Method of Obtaining Standard of Safety

The regulations administered by the CPSC cover the safety aspects of products such as flammable fabrics, hazardous household substances, toys, and children's articles. Provision was made to obtain a certificate of safety for products that meet CPSC standards. Congress noted that anyone can petition the CPSC for the issuance of a safety regulation. A response to the petition must be made and if there is a need for a regulation, certain procedures must be followed.

A notice should be published in the *Federal Register* stating the product and risk involved. The CPSC must state its intention to regulate the specific item and must inform the public about any existing standards. Readers and other interested parties then have 30 days to submit standards for CPSC perusal or help the commission create new standards. In addition, the CPSC is specifically prohibited from recommending an individual package design for the product. All standards must consider the state of the art of present day packaging technology and are to be basically created by a task force. If a standard of safety cannot be formulated, the CPSC does have the ability to effectively ban the hazardous product. This was done in late 1972 when detergents containing over 10% hydroxides were banned from the market.

Prior to the issuance of a standard of safety, the CPSC does require supporting test data and voluntary assistance from consumer groups. A complete written record is available for public scrutiny and all meetings are open to the public.

If, after a standard of safety is issued, there are objections, judicial review is open to the objector. The CPSC may also ask the U.S. District Court to seize a product if it is "an imminently hazardous consumer product." Any product which is regulated must be certified as to its conformance to the standard of safety.

Certificates are usually placed on the label of household chemicals or for toys may be inserted in the box stating the products' conformance to possible electrical, mechanical and/or thermal standards of safety. Toys may also require a permanent certification label noting the date, product manufacturer, and address.

The testing program for certification must be acceptable to the CPSC. If the CPSC feels that there is an element of doubt as to the ability of the manufacturer to test the product, they may request a third party such as the Consumers Union to perform the appropriate tests. Non-compliance may involve several courses of action. A public hearing may be held to determine the hazard involved. A public notice may be inserted in newspapers about the hazard or a notice may be mailed to all distributors of the product. If the hazard involved is more serious, the questionable part may be replaced or repaired on request by registered letter to the purchaser. The last resort is a total recall notice of the product. This did occur around Christmas time with certain hazardous imported toys. In addition, the CPSC does have the ability to enter and inspect a plant and full records must be made available for their inspection.

An imported item covered by the CPSC must have a certificate prior to entry into the U.S. If it proves to be hazardous after entry, it may be seized. The CPSC also maintains close liason with foreign regulatory bodies and if they are aware that the foreign manufacturer does not keep proper records (few do) and the product is hazardous, they may request the Treasury Department to stop entry of the product into the U.S.

These laws are extremely rigid for imports since they were specifically designed to prevent the entry of hazardous toys into the U.S. There are no regulations covering exports into foreign nations.

Penalties for non-compliance range from $2000 (civil) to $50,000 and/or one year in jail (criminal). The 1972 regulation also established the ability of the injured party to sue, with certain limitations, in Federal Court.

Specific Regulations

The FFA of 1953 (amended in 1967) covers items such as garments, bedding, and drapery. It sets labeling requirements for these products and has been notably useful in certain areas. Flammable fabrics were prohibited from being used in baby nightwear and also prohibited from use in tent material.

The FHSA of 1960 requires that product labels carry information to warn consumers against the dangers of hazardous materials that are intended for home use, or that might be expected to find their way into the home. This law does not apply to industrial products intended for industrial use. Excluded were products subject to the Atomic Energy Act and fuels and their containers. This act effectively repealed the *Caustic Poison Act* of 1921 except for dangerous caustic or corrosive materials subject to FDA jurisdiction. These products are mainly hair preparations.

Poison Prevention Packaging Act — 1970.—The PPPA of 1970 was introduced by Senator Moss of Utah in 1969. It was proposed as an amendment to the FHSA of 1960 and was restructured in 1970 prior to passage. The bill introduced was designed to prevent children under 5 years of age "to obtain a toxic amount of product in a reasonable time." While a large number of children under 5 are poisoned every year, it is doubtful whether safety closures and/or child-resistant flexible packaging are the total answer. There is certainly no substitute for an inaccessible storage place for dangerous containers.

Assisted by a technical advisory committee and a study conducted by Dr. Wilton Krogman, a testing procedure (protocol) for special packaging was established for the PPPA.

This is composed of a panel of 200 children used to test the effectiveness of various safety closure designs. The children are between 3½ and 4¼ years old. This group, which is three times larger than the minimum number needed to produce significant statistical results, is given 10 min (two 5-min periods) to open the closures. The children are even paired off to include the possibility of a "team attack" by the youngsters.

Additionally, a panel of 100 adults (70% female) is used, age 18 to 45 years inclusive.

Dr. Robert G. Scherz, chief of pediatrics service for Madigan General Hospital, proposed an alternative testing protocol for safety packaging. Alternative Protocol, Dr. Scherz's, procedure reduces the number of children in the test panel from 200 to 100 but required all 100 to be male.

Courtesy of Sen. Moss' Office

FIG. 5.3. SEN. FRANK E. MOSS
Introduced PPPA of 1970 in Congress.

The adult panel would be reduced from 100 (70 females and 30 males) to 70, all females.

The technical advisory committee reviewed these procedures and selected the "Press Protocol." This test was published in the *Federal Register.* The CPSC is now emphasizing adult resecuring of child resistant closures so as to maintain their safety feature. In a recent case against Miles Laboratories, the closure on its vitamins failed the test as to adult resecuring. A cap change ensued and all were satisfied.

Courtesy of Government of the Netherlands

FIG. 5.4. DUTCH "CHILD WELFARE STAMP"
Issued as a set of four stamps, the 45+20¢ stamp deals with the risk of poisoning when a child has access to drugs, detergents, and other substances kept in medicine cabinets and kitchen cupboards. The surcharge benefits institutions active in the field of child welfare.

Substances that require safety closures to protect children from serious personal injury or serious illness are: (See Table 5.1 for testing companies.)

(1) Aspirin — any aspirin containing preparation for human use in a dosage form intended for oral administration (August 14, 1972);[1]

(2) Furniture Polish — nonemulsion type liquid furniture polishes containing 10% or more of mineral seal oil and/or petroleum distillates (September 13, 1972);

(3) Methyl Salicylate — liquid preparations containing more than 5% by weight of methyl salicylate (September 21, 1972);

(4) Controlled Drugs — any preparation for human use that consists in whole or in part of any substance subject to control under the Comprehensive Drug Abuse Prevention and Control Act of 1970 (October 24, 1972);

(5) Sodium and/or Potassium hydroxide — household substances in dry forms such as granules, powder, and flakes, containing 10% or more by weight of free or chemically unneutralized sodium and/or potassium hydroxide, and all other household substances containing 2% or more by weight of free or chemically unneutralized sodium and/or potassium hydroxide (April 11, 1973);

(6) Turpentine — household substances in liquid form containing 10% or more by weight of turpentine (July 1, 1973);

(7) Methyl Alcohol — household substances in liquid form containing 4% or more by weight of methyl alcohol (July 1, 1973);

(8) Sulfuric Acid — household substances containing 10% or more by weight of sulfuric acid, except such substances in wet cell storage batteries. (August 14, 1973);

(9) Kindling and/or Illuminating Preparations — prepackaged liquid kindling and/or illuminating preparations which contain 10% or more by weight of petroleum distillates (October 29, 1973);

(10) Prescription Drugs — any drug for human use that is in a dosage form intended for oral administration and that is required by Federal law to be dispensed only by or upon an oral or written prescription of a practitioner licensed by law or administer such drug (April 16, 1972);

---

[1] The date given is the effective date or date proposed for packaging standards as published in the *Federal Register*.

(11) Economic Poisons — household substances that are economic poisons that are packaged in liquid forms in less than one-gal. containers or in nonliquid forms in packages of 5 lbs or less, and that have been found to be highly toxic by human experience (September 14, 1972);

(12) Ethylene Glycol — household substances in liquid form containing 10% or more by weight of ethylene glycol. Includes automotive antifreeze and certain automotive brake fluids (December 28, 1972);

(13) Paint Solvents — solvents for paint or other similar surface-coating material solvents, such as thinners, removers, and brush cleaners in liquid form which contain 10% or more by weight of petroleum distillates, benzene, toluene, xylene, or combinations thereof (February 9, 1973);

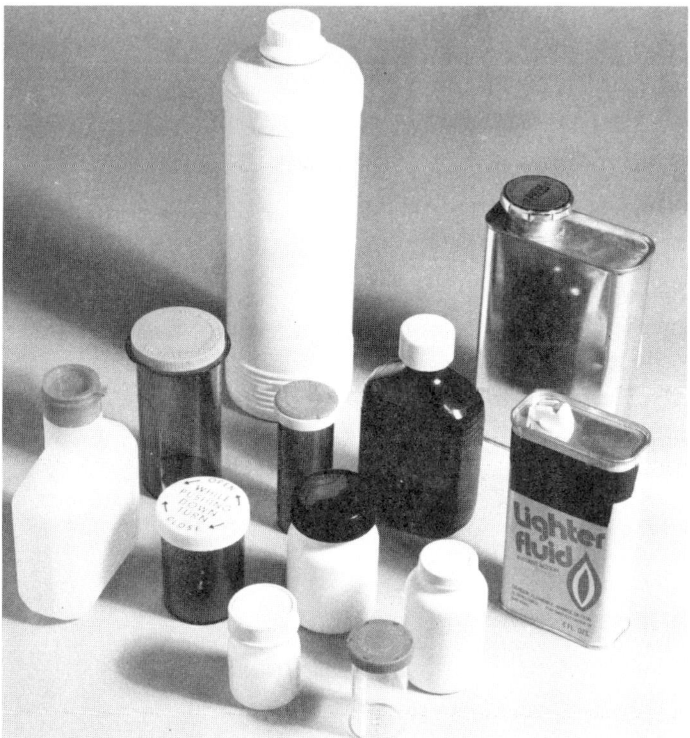

Courtesy of Consumer Product Safety Commission

FIG. 5.5. PRODUCTS IN CHILD RESISTANT CLOSURES
These products illustrate a wide range of products in child-resistant closures.

TABLE 5.1. MAJOR TESTING COMPANIES FOR CHILD-RESISTANCE REGULATIONS

Affiliated Medical Enterprises (AME)
Princeton Pike
P.O. Box 57
Princeton, N.J. 08540
Telephone: (609) 924-9658

Allied Chemical Corporation
Specialty Chemicals Division
P.O. Box 1057 R
Morristown, N.J. 07960
Telephone: (201) 538-8000

American Institutes for Research
Accident Research Center
8555 Sixteenth Street
Silver Spring, Maryland 20910
Telephone: (301) 587-8201

Associated Testing Laboratories
9 Brighton Road
Clifton, N.J.
Telephone: (201) 473-6455

Burke Marketing Research, Inc.
3220 North Main
Dayton, Ohio 45405
Telephone: (513) 277-3184

Child-Related Research, Inc.
4565 Fortuna Way
Salt Lake City, Utah 84117
Telephone: (801) 277-7982

Cornell Aeronautical Laboratory
Center for Consumer Research
P.O. Box 235
Buffalo, N.Y.
Telephone: (716) 632-7500

Foster D. Snell, Inc.
Division of Booz, Allen and Hamilton, Inc.
General Laboratories
Hanover Road
Florham Park, N.J. 07932
Telephone: (201) 377-6700

Franklin Institute Research Laboratories
Twentieth and Parkway
Philadelphia, Pennsylvania 19103
Telephone: (215) 448-1388

Grant Morgan Associates
405 N.W. 18th Avenue
Portland, Oregon 97209
Telephone: (503) 222-2374

Hill Top Research, Inc.
Miamiville, Ohio 45147
Telephone: (513) 831-3114

Home Arts Guild
Research Center
29 West Randolph Street
Chicago, Illinois 60601
Telephone: (312) 726-7406

IIT Research Institute
Engineering Mechanics Division
10 West 35th Street
Chicago, Illinois 60616
Telephone: (312) 225-9630

University of Maine
Orono, Maine 04473
Telephone: (207) 947-4591
(University Contracting Office, Bangor, Maine)

Robert G. Scherz, Col., M.D.
Madigan General Hospital
Tacoma, Washington 98431
Telephone: (206) 967-6743

United States Testing Co.
Consumer Research Division
1415 Park Avenue
Hoboken, N.J. 07030
Telephone: (201) 792-2400

Western Technical Associates
4940 West 147th Street
Hawthorne, Calif. 90250
Telephone: (213) 679-9719

(14) Promotionally-Distributed Samples — promotional samples of any household substance distributed directly to the household but not presented in person to a responsible adult member thereof, and for which any cautionary labeling is either required or recommended, shall, except for substances in pressurized spray containers where the only hazard is that the contents are under pressure (February 9, 1973);
(15) Acetaminophen containing products (August, 1978);
(16) Iron containing drugs and dietary supplements containing 250 mg of iron or more (August, 1978);
(17) Over-the-counter (O.T.C.) antihistimines.

The manufacturer or packer of any of the substances listed must provide the Commissioner with a sample of each type of safety packaging used, as well as the labeling for each size of the product and any noncomplying package.

The FDA has also set rules for noncomplying packages permitted under the PPPA. Manufacturers must meet five conditions to market products in noncomplying packages intended for the elderly and handicapped. The conditions are: closures must be difficult for these people to use; child-resistant packaging is on popular sizes of the product; packages with safety closures are adequately distributed and advertised; the noncomplying size is the one most likely to be used by the elderly; and the exempt package bears required label warnings.

Future.—Total consumer protection is probably an unrealistic dream with costs resulting in hundreds of millions of dollars. Coupled with this cost would be a bureaucracy that would be intolerable to the public. The organization of the CPSC has also been a problem. Meetings are open to consumerists and they become cumbersome and difficult to manage. Many child resistant closures do not function properly and several standards are too difficult to meet with present technology.

There are efforts underway calling for consolidating the CPSC, FDA, EPA, and Department of Agriculture into a single agency, either separate from the HEW or at a higher level within the department. There is also concern that this in itself might create a bureaucracy that would be more inefficient and difficult to administer than the FDA.

A balance must be made between the financial pressures of consumer protection and the possible resulting layoffs and closings of polluting factories.

Aerosol Containers, U.S.— The Food and Drug Administration recently announced that it would require a warning label on aerosol containers using fluorocarbon propellants as a first step toward eventual elimination of virtually all such products.

The Consumer Product Safety Commission said in a joint announcement with the FDA that it was proposing similar action for the products it regulates.

The actions came in response to a National Academy of Sciences report last year stating that the propellants formally known as chlorofluorocarbons rise into the stratosphere and deplete the earth's ozone layer, a thin protective gaseous belt that filters harmful ultraviolet radiation from the sun.

The FDA said that all food, drug and cosmetic containers using fluorocarbon propellants and shipped in interstate commerce must carry the following warning after October 31, 1977:

> "Warning: contains a chlorofluorocarbon that may harm the public health and environment by reducing ozone in the upper atmosphere."

About 85% of aerosol products containing the chemicals are covered by the FDA regulation. The 15% remaining are regulated by the Consumer Product Safety Commission or the Environmental Protection Agency.

EPA already requires a warning label on pesticide aerosols containing fluorocarbons, and the product safety commission proposal is expected to take effect early next year.

The joint announcement stated that details of a mandatory phase-out program for all nonessential aerosols using fluorocarbons would be announced soon.

The Food and Drug Administration also said that its order would not apply to over-the-counter drugs used for inhalation therapy for bronchial asthma, contraceptive vaginal foams and cytology fixatives that are medical devices used in a cancer diagnostic procedure. Use of fluorocarbons in those products is essential, the agency said.

Aerosol Containers — Europe.— A Swedish government action of international packaging importance is a recent decision by Swedish product control authorities to ban all imports of aerosols containing the chlorofluorocarbon "Freon" as a propellant, with the exception of certain medical products.

This rule will go into effect January 1, 1979, and joins government action elsewhere, particularly in the United States, to eliminate fluorocarbons as propellants for most pressure-packaged products because of suspected health and ozone-damaging threats. Similarly, Swedish foodstuffs authorities have moved against polyvinyl chloride (PVC) when used in food packaging, another subject of government action elsewhere. elsewhere.

As of July 1, 1978, the vinyl chloride monomer (VCM) level in foodstuffs packed in PVC must not be higher than 0.01 mg/kg, compared with 0.05 now. The same authorities propose to lower the limit for tin (pewter) found in foods from 250 to 150 mg/kg.

Soft Drink Bottles.— In 1976, an estimated 125,000 Americans were injured by glass from carbonated soft-drink bottles that exploded on shelves in normal handling or when jarred.

Many of those injured required stitches and other treatment in doctors' offices and emergency rooms, and still others required hospitalization and special treatment to repair damaged arteries.

It is not known whether any deaths have resulted from exploding soft-drink bottles. Nor is there an accurate statistical breakdown according to soft-drink brands or size and type of bottle because collection of such data is voluntary and incomplete. Further, the reasons for the explosions are not clear. In an attempt to make soft-drink bottles of more uniform, stronger quality, however, the Glass Packaging Institute, the National Soft Drinks Associations and the National Bureau of Standards, an agency of the Department of Commerce, recently agreed on the new standards. The action was prompted in part by lawsuits against the industry.

The standards, which apply to both returnable and disposable glass bottles, set levels for the pressure strength of bottle walls and for the ability of a bottle to withstand sudden changes in temperature such as occur when it is removed from a refrigerator on a hot summer day.

A group of doctors in Phoenix reported in the Journal of the American Medical Association on their experiences in treating patients who had been injured by shattering glass from exploding bottles. In one case, a 7-month-old infant grasped a two-qt pop bottle that exploded, severing an artery and nerve in the child's arm and damaging muscles. the infant required extensive surgery and blood transfusions. Other patients lost eyes when glass propelled by an explosion pierced their eyelids and imbedded in the eye.

It is not known how many of the bottles now in circulation meet the new industry standards. Also, there is no way that a consumer can distinguish between bottles made before and after adoption of the new standards. Accordingly, safety experts advised consumers to observe the suggestions made by the Arizona doctors, which included:

(1) Storing bottles in a cool place and placing them on the floor or lowest shelves to reduce the hazard in case the bottle explodes;
(2) Avoiding jostling bottles or hitting them together;
(3) Directing the cap away from the body or face when opening the bottle.

Bottles that contain carbonated drinks are believed to be more hazardous than those filled with non-carbonated liquids because of the greater internal pressure. Consumers are advised to refrigerate bottles before use to reduce the internal pressure.

The Arizona doctors said that returnable bottles are thicker, heavier, and may withstand greater internal pressure than the lighter-weight, non-returnable bottle. As they are reused, however, bottles may become cracked, scratched or weakened. No structural tests are run by bottlers and only brief inspection is performed. There is wide variation in the estimates of the number of times that returnable bottles are recycled. The figures range up to 16 times. Also, the doctors cited concern that the larger 32 to 64-oz bottles are more dangerous in an explosion than smaller ones because fragments are propelled with greater force.

The National Soft Drinks Association says that the new standards, although voluntary, will be used as "a purchasing tool" when the 2135 soft-drink manufacturers contract with the 90 companies that make the glass bottles.

According to the standards, a bottle's side walls must be able to withstand pressure of 225/sq in. for returnable bottles and 200 lbs for disposable bottles. Previously, there was no such recommended figure for either type of bottle.

Also, the standards set a minimal level for sudden changes in temperature, called thermal shock. New bottles will be manufactured to withstand the thermal shock when transferred from a hot-water bath of 62.8°C (145°F) to a colder bath of 21.1°C (70°F).

INDUSTRIAL

Industrial materials are defined as those products which are sold and shipped by one industrial concern to another and are not intended to be used in one's home. Those materials considered as "hazardous" will be covered from a regulatory viewpoint. Included are products such as *explosives, corrosive liquids, compressed gases, flammable liquids and solids,* and additional items deemed hazardous to handle and transport.

In 1866, Sen. Chandler of Michigan introduced a bill designed to prohibit the shipping of nitroglycerine in any type of vehicle used to transport passengers. From this beginning piece of legislation evolved many separate and distinctive regulations dealing with hazardous substances. Independent codes were formulated over the years by many different agencies. Between 1970-1960, The Coast Guard had jurisdiction over hazardous substances shipped by water; the Federal Aviation

Agency (FAA) had control over transport by air; the Interstate Commerce Commission (ICC) had jurisdiction over rail and truck shipments and worked closely with groups such as the Association of American Railroads and the American Trucking Association. Flammable products were tested by standards developed by the N.Y. Fire Department. The Bureau of Mines covered explosive regulatory affairs.

While all these agencies established workable regulations, the need for a unified law became apparent. In 1966, Public Law 89-670 established the Department of Transportation (DOT) which was given the regulatory authority relative to the shipment and transportation of hazardous materials by all civil modes of transportation. The DOT act specified that the responsibility for water safety be the Coast Guard's; air safety was the responsibility of the FAA; railroad safety was handled by the Federal Railroad Administrator; and the Federal Highway Administrator covered motor carrier safety. There was still a definite need for the establishment of one group capable of administering all the various regulations dealing with hazardous materials transport. This was accomplished by the creation of the Hazardous Materials Regulations Board (HMRB).

Hazardous Materials Regulations Board

Created in 1967, the HMRB provided the information necessary for the identification of hazards during shipment by various methods. (See Table 5.2 for shipping methods.) They defined hazardous materials and separated them into various classes. These are:

(1) Explosives
(2) Flammable liquids,
(3) Combustible liquids,
(4) Corrosive materials,
(5) Poisons,
(6) Radioactive materials,
(7) Etiologic agents,
(8) Oxidizing materials,
(9) Flammable solids,
(10) Compressed gases,
(11) Other restricted articles, and
(12) Hazardous articles.

Not covered by the DOT-HMRB are pesticides, foods, drugs, cosmetics, fuels, and radioactive materials already covered under AEC regulations. In cases where identification of a certain product is doubtful, the shipper had to consult the HMRB or the Bureau of Explosives. Each class of hazard specified above was covered by different regulations specifying their particular problem. Examples of typical definitions used by the HMRB follow;

(1) *Corrosive Material* — "A corrosive material is a liquid or solid that causes visible destruction or irreversible alteractions in human skin tissue at the site of contact, or in the case of leakage from its packaging, a liquid that has a severe corrosion rate on other material such as steel or aluminum."

(2) *Oxidizing Material* — "An oxidizing material is a substance such as a chlorate, permanganate, peroxide, nitrocarbo nitrate, or a nitrate that yields oxygen readily to stimulate the combustion of organic matter."

(3) *Compressed Gases* — "A compressed gas is defined as any material or mixture having in the container an absolute pressure exceeding 40 p.s.c. at 70°F, or regardless of the pressure at 70°F, having an absolute pressure exceeding 104 p.s.c. at 130°F, or any liquid flammable material having a vapor pressure exceeding 40 p.s.c. absolute at 100°F as determined by ASTM Test D-323."

Other definitions were as comprehensive as those mentioned above.

TABLE 5.2. SHIPMENT OF HAZARDOUS MATERIALS — OTHER METHODS

Method	Organization	Regulation
Mail	U.S. Code, Chapter 83 U.S. Postal Service	Title 39, Parts 123 & 124 Title 49
Private Carriers Contract Carriers	Interstate Commerce Act	49 CFR 177
Bus	National Bus Traffic Association	Rule 16, Section B, Northeastern Express Tariff No. A-603-C
United Parcel Service	United Parcel Service	Guide (in text)

For flammable liquids, the definition is "a flammable liquid is any liquid which gives off flammable vapors (as determined by flash point from Tagliabue's open-cup tester, as used for test of burning oils) at or below a temperature of 80°F." This test consists of placing the substance into an open metal cup and gradually raising the temperature until a lighted match will ignite the vapors in the cup. Flammable liquids are categorized by the open-cup test as:

(1) Flash Point below −6.7°C (20°F) —*extremely flammable*;
(2) Flash Point above −6.7° (20°F)–26.7°C (80°F) — *flammable;*
(3) Flash Point between 26.7°C (80°F)-65.6°C (150°F) — *combustible.*

There has been considerable discussion about the relative merits of an open-cup test vs. a closed-cup (Abel or Abel-Pensky) test. While the open-cup test allows for the continuous escape of vapor and is better for mixed compounds,the closed-cup test is more reproducible and concentrates the vapor yielding a lower flash point. A 26.7°C (80°F) value in an open-cup test is about equal to a 22.8°C (73°F) value in a closed-cup test. There is very little correlation between both test procedures; however, the open-cup method is standard everywhere in the U.S. except for North Dakota which has a special point law on its books and specifies the closed-cup test for flammable liquid determination and classification.

Materials Transport Bureau

In 1976 new legislation was introduced that established the Materials Transport Bureau (MTB) published in the *Federal Register* on April 15, 1976, as HM-112. The MTB was given the authority to coordinate all separate codes, remove the ones not deemed necessary and recodify all the basic parts of the law. Administration of the MTB is the responsibility of the DOT under whose jurisdiction the MTB operates.

Initially the MTB set up the consolidation of all hazardous materials regulations. In part 1 of the 1976 legislation, the DOT stated that they recognized the inherent differences between shipping hazardous materials by air, sea, and land and would recognize these variations in this new legislation. Once again, all types of hazardous materials were defined. In Section 172, a truly comprehensive listing of all hazardous materials is found. Tabulated over many pages in HM-112, all hazardous substances are listed and classified as to type, DOT Classification, labels required, exceptions, packaging required, maximum shipping quantity, water shipments, and other requirements. This list obsoletes all previous regulations and gives both the codes and reference numbers for each product to be looked up in later portions of HM-112. For example, under package requirements, the table might state 173.153. This can be looked up in later pages and the type of package needed for the particular product would be fully described.

HM-112 also consolidates the various color codes, type sizes for labels and labeling required.

Courtesy of American Trucking Association

FIG. 5.6. FAULTY PACKAGING CAUSES SPILLAGE
Such hazardous materials must be properly contained in order to prevent damage due to shipping.

Section 173 lists all materials by their hazardous substance classification and the packages required for the item. At the end of 173 is a section on the placards that are necessary to be placed on the outside of trucks or railcars for certain products. Exactly how they are to be placed and the colors needed are specifically mentioned. Section 174 deals with the definition of various materials. Explosives are defined as Class A (detonating or otherwise of maximum hazard), Class B (which in general function by rapid combustion rather than by detonation) and Class C (certain types of manufactured articles which contain Class A, or Class B explosives, or both, as components but in restricted quantities, and certain types of fireworks). Poisons are defined: Class A (poisonous gases or liquids of such nature that a very small amount of the gas, or vapor of the liquid, mixed with air is dangerous to life); Class B (those substances, liquid or solid [including pastes and semi-solids] other than Class A or Class C poisons, which are known to be so toxic to man as to afford a hazard to health during transportation, or which, in the absence of adequate data on human toxicity, are presumed to be toxic to man because they fall within "certain laboratory test result ranges as listed in 49 CFR

173.343 and in 46 CFR 146.25-10"); and Class C (tear gases or similar irritating substances. General handling and loading requirements are mentioned for all hazardous materials.

The DOT has stated that it will take about five years to codify all hazardous materials and fully reorganize all the classifications. Also, the 1976 legislation obsoletes all previous regulations.

United Parcel Service (UPS) and Others

For shipment by truck, the United Parcel Service (UPS) has published a well organized manual which lists all hazardous materials and correlates the DOT information for land shipment with UPS requirements. (See Table 5.3.) In most cases, UPS standards are more severe than those of DOT. The UPS manual (with annual supplements) is available from:

United Parcel Service
51 Weaver Street
Greenwich Office Park 5
Greenwich, Ct. 06830

Courtesy of United Parcel Service

FIG. 5.7. UNITED PARCEL SERVICE
View of UPS central shipping facility.

TABLE 5.3. REPRESENTATIVE PAGE FROM UPS GUIDE

Hazardous Material (and Chemical Formula)	Required DOT Shipping Name	Classification, Hazard Information No. and DOT Label Required if Not Excepted	Package Number	Maximum Unit Quanity
Spent iron sponges, See Iron sponge, spent	Iron sponge, spent	Flammable solid, 40, Flammable solid		Not Accepted
Spent mixed acid. See Nitrating acid, spent	Nitrating acid, spent	Corrosive material, 51, Corrosive	24	1 Liter
Spent oxide, See Oxide, spent	Oxide, spent	Flammable solid, 40, Flammable solid		Not Accepted
Spent sulfuric acid. See Sulfuric acid, spent	Sulfuric acid	Corrosive material, 84, Corrosive	24	1 Liter
Spirits of nitrogylcerin (1 to 10%) $NO_3\ CH_2\ CH(NO_3)\ CH_2\ NO_3$	Spirits of nitrogylcerin	Flammable liquid, 34 Flammable liquid		Not Accepted
Spirits of nitroglycerin, not exceeding 1% nitroglycerin by weight $NO_3\ CH_2\ CH(NO_3)\ CH_2\ NO_3$	Spirits of nitroglycerin not exceeding 1% nitroglycerin by weight	Flammable liquid, 30 Flammable liquid (E)	24	1 Liter
*Spirits of salt, See Hydrochloric acid	Spirits of salt or Hydrochloric acid	Corrosive material, 80, Corrosive (E)	14A 25 33	18.93 Liters
Sporting powder. See Black powder or Propellant explosive, solid, Class B Explosive	Black powder or Propellant explosive, solid, Class B explosive	Class B explosive 17, Explosive B		Not Accepted
Spray starting fluid. See Engine starting fluid	Engine starting fluid	Flammable gas, 23, Flammable gas	8	1 Kilogram
Spreader cartridge. See Fireworks, special	Fireworks, special	Class B explosive, 17, Explosive B		Not Accepted
Squib, electric or safety. See Electric squib or safety squib	Electric squib or Safety squib	Class C explosive, 15, Explosive C	15A	22.5 Kilograms
*Stain, See Paint, enamel, lacquer, stain, shellac, varnish, etc.	Not regulated in UPS ground service	Combustible liquid, 30, None	22 17	18.93 Liters subject to 22.68 Kilograms gross weight limit

Courtesy, United Parcel Service

The Manufacturing Chemists Association (MCA) has issued a "Guide to Precautionary Labeling of Hazardous Chemicals", commonly known as the LAPI Manual. It is an authoritative and useful publication which is of great value in the preparation of precautionary data for any type of material governed by the MTB. The LAPI Manual covers hazardous materials in a broader system than the UPS manual, but is somewhat more difficult to read.

The Chemical Specialties Manufacturers Association (CSMA) has published several brochures:

(1) "Compilation of Labeling Laws for Hazardous Products"
(2) "Compilation of Labeling Laws for Antifreeze and Brake Fluids"
(3) "Aerosol Guide"
(4) "Laws and Regulations of Interest to the Aerosol Industry"

Other organizations that have published useful data on hazardous materials include the National Paint and Varnish Association (NPVA), the National Fire Protection Association (NFPA), the Compressed Gas Association (CGA) and the Aluminum Association. Organizations such as these, and others, are valuable sources of information on the handling, packaging and storing of hazardous substances.

Practical Aspects

The practical implications of shipping hazardous materials are manifold. Industrial samples, new products and commercial shipments are part of the types of materials covered by hazardous material regulations. A new solvent, shipped to a customer, must also conform with the proper regulations.

An interesting case study concerns Fisher Scientific, a large and well-established laboratory supply house headquartered in New York City. During the 1970's, Fisher was awarded a contract to supply Uganda with sufficient laboratory equipment for 120 new teachers' colleges. The contract included all the materials, hardware, and chemicals necessary for 120 laboratories. Various airlines were chartered to airlift the materials to Uganda. Although certain products were considered too deadly to transport by air (carbon disulfide was shipped by sea), the major problem faced by Fisher was how to meet the many regulations involved in transmitting the materials from their New Jersey warehouse to Kennedy Airport in New York City. In crossing state lines, Fisher was faced with meeting a myriad of regulations, i.e., federal, state, city, tunnels, bridges, airport, etc., for over 1500 different materials. The problems were successfully solved by the creation of five manuals on labeling by Fisher and also labeling the various products with all that was necessary to conform with the appropriate regulations. In addition, the type of packaging required was extremely conservative. For example, an acid was packaged in a bottle within a slip-cap metal can filled with vermiculite. The can was then taped and imbedded in a wooden crate filled again with vermiculite. Enough vermiculite was used to absorb the total amount of acid if breakage occurred. Other regulations were rigid as those for acid packaging.

It should be noted that the shipment to Uganda was successfully handled.

International Standardization

The International Standards Organization (ISO) has a subcommittee which seeks to establish worldwide hazardous materials regulations. At present, the DOT does participate in subcommittee meetings by sending one delegate to the meetings, which are usually held in Europe for 2-3 weeks duration. The problem facing the DOT is their financial limitation as to delegate participation. The U.S. is thus in an unfavorable position and the European nations have created various tariffs, and regulations favoring themselves and unfavorable to U.S. industry. Even though over $3 billion of hazardous materials are exported to Europe by the U.S., it is becoming more and more difficult to maintain this margin every year. The DOT is also very careful about funding industrial experts because of possible consumer complaints. The net result is that Europe appears to be limiting U.S. exports of hazardous substances.

REFERENCES

ANON. 1974. EPA seeks input from SCMA on impact of child resistant packaging regs. *3,* (4)13-14.

ANON. 1975A. Standard package said next consumer battle. Food & Drug Pkg. *32,* (11) 3, 19.

ANON. 1975B. CPSC outlines its jurisdictions over aerosols regulated by FDA. Food Chem. News *17,* (20) 16-18.

ANON. 1976A. CPSC answers GAO's charges of lax, ineffective enforcement. Food and Drug Pkg. *35,* (8) 10.

ANON. 1976B. GAO tells CPSC: "Clean up your acts." Mod. Pkg. *49,* (10) 17,20.

ANON. 1976C. Poison prevention packaging. Fact sheet No. 46. U.S. Product Safety Commission, 3pp.

CUNNINGHAM, W.H. 1976. Consumer protection: More information or more regulation? J. of Mkg. *40,* (2) 63-69.

HICKOX, R. 1976. GAO blasts CPSC efforts at enforcing safety regs. but asks greater authority. Food & Drug Pkg. *35,* (5) 1, 24.

HICKOX, R. 1976. CPSC seen asking CR closures for cleaners with 2% ammonia. Food & Drug Pkg. *35,* (3) 4, 29.

HICKOX, R. 1977. CPSC staff asks reevaluation of agency's PPPA activities. Food & Drug Pkg. *37,* (7) 4, 29-30.

LEHMAN, H.H. 1975. Safety packaging preventing poisoning. J. Envir. Health. *38,* (2) 108.

MARDER, H. 1975. A breather on regulations. Mod. Pkg. *48,* (2) 48, 50.

SALWASSER, M.L. 1977. Free enterprise threatened by government bureaucrats. Good Pkg. *38,* (3) 12-13.

SIMPON, R.O. 1975. CPSC chairman views PPPA progress. Food & Drug Pkg. *32,* (12) 12-13, 25.

STEVAN, M.A. What shoppers are concerned about. FDA Consumer *8,* (7) 13-17.

6

State, Local, Federal and Military Regulations

STATE

Traditionally, the areas reserved for state involvement with packaging regulatory affairs have been in the fields of weights/measures and local fresh products. It has only been within the last 20-25 years that states have become totally involved with packaging regulations. Their participation in this field has largely been the result of the affluent 1945-1960 period.

In the post World War II years, many new products and even newer packaging appeared on the market. States and municipalities spent various amounts of money on social welfare and schools but almost totally ignored the mounting trash accumulation and environmental matters. Toward the end of the 1950's, a groundswell appeared on the subject of the environment. Many of the older trash dumpings were closed. Incinerators were closely checked and the many inefficient ones were rehauled and if not useful, put out of action. It was these two factors that evoked one of the earliest pieces of local legislation dealing with packaging.

In the early 1960's, New York City, under the Lindsay Administration, maintained an agency headed by Jerome Kretschmer, that dealt with environmental matters. Because of Mr. Kretschmer's concern about packaging waste and the fact that the disposal of municipal trash had become a vexing problem, New York City attempted to pass a law that would place a tax on packaging proportional to the disposal problems inherent in the specific material. The materials were rated as low, medium, and high. The bill was defeated and held unconstitutional because it discriminated between the different materials. But the problems dealing with the disposal of municipal trash had not been solved. Even though waste due to packaging is not a large percentage of the total national waste, it can form a large percentage (up to 35%) of municipal urban waste. In rural areas, trash due to packaging is a much less important factor in the total environmental cycle.

This problem was compounded by an even stronger national cry stating that full product visibility was necessary for all packages and that the consuming public was being hoodwinked because not all sides of a package could be viewed prior to sale.

The cudgel for "full visibility" was taken up mainly by local officials and many debates followed between the opaque tray industry (pulp and expanded polystyrene foam) and the clear polystyrene tray interests. In one case, there was almost total confusion relative to the pertinent issue.

A New Jersey community passed a regulation stating that all refrigerated products *must be* in totally transparent containers. Discussed on a Friday night and then passed by the City Fathers, the town found itself deluged with packaging and food personnel the next day. If not repealed (and it was *rapidly* on Monday), the town's citizens would not be able to purchase milk, cheese or the many other refrigerated products on a regular basis. But the present trends had already started. As a result, Florida now requires total visibility on all fresh meat and produce packages.

States, cities, and towns had become more directly involved in packaging regulations. And joining the cry for the proper disposal of municipal waste were the increasing attacks against the one-way beverage container.

As a total factor, municipal waste is not the largest waste problem facing the civilized world. Human, agricultural, and mining wastes are even bigger problems in the environment. The important point about municipal waste (50-60 million tons/yr) is that it does contain a large amount of recyclable items. This caused industrial personnel to organize the *National Center for Resource Recovery.* Aimed at being a clearing house for all regulations and, subsequent to the passage of the *Resource Recovery Act of 1970,* a coordinating body for the federal seed money intended to fund research on waste recovery and sorting, it has served as a vital factor in the recovery area. Several projects are now underway. American Can Company maintains a pilot project on waste sorting in Milwaukee and both the General Electric Company and Black Clawson have methods available to efficiently deal with waste recovery.

As with all projects, localities face a difficult choice. Should the total waste effort be aimed at landfill or recovery? While San Francisco prefers landfill, New York City needs recovery. Materials are also a factor in a proper selection. Certain packaging can be recycled economically and others cannot. Both aluminum and steel cans are successfully being recycled. Based on voluntary projects, about 25% of all aluminum cans are being recycled. And it takes less energy to reprocess aluminum cans than to produce virgin aluminum. About 22% of all paper is also being recycled. And glass is also heavily into recycling. It has been found that a furnace can be fired with up to 35-40% of broken glass.

The stage has thus been set by all the above factors. And now many packaging personnel believe that the greatest threat to the industry will be in the area of future state and local regulations. Fully informed, intelligent and proper regulations are not a threat. The difficulty lies in the passage of well-meaning but foolish regulations that could effectively destroy an industry and indirectly affect a nation's economy.

Legislative Attempts to Control Packaging

With only a few recent exceptions, legislative attempts at the state and federal levels have been directed at the nonrefillable beer and soft drink container. Public recognition of the solid waste problem had already been growing since the early 1960's. This group action culminated in Congressional legislation with the passage of the *Solid Waste Disposal Act of 1965.* Here was the first piece of federal legislation that tried to reduce environmental hazards created from the mounting piles of solid waste. Succeeded by the *Resource Recovery Act of 1970,* the climate created in Washington clearly was favorable to legislation that would be, in some degree, restrictive to packaging.

In 1972, the Environmental Protection Agency (EPA) published a study that carefully examined packaging and investigated the cost of each material as a measure of the total energy expanded. A flow chart, incorporating seven steps, was designed to serve as an analytical tool in package development. Each step measured the energy spent vs. results obtained for a specific material. These original EPA studies have now become the reference point for all present day energy considerations in package material production.

The action created by both the passage of the *Solid Waste Disposal Act of 1965* and the release of the EPA study caused many state legislators to react quickly to the subject of solid waste and to packaging's contribution to this problem. Although over 1000 state-wide measures to control packaging on some level were proposed, only seven states have acted and passed restrictive packaging legislation. These states (Oregon, Vermont, Maine, Michigan, Iowa, Connecticut and Delaware) have not banned the one-way beverage container. They have imposed various restrictive deposits on these packages and have, in effect, caused a serious erosion in the market for one-way containers in those states.

On July 2, 1972, Governer Tom McCall of Oregon signed into law a two-tier bill that became effective in October, 1972. (See Table 6.1.) It called for a minimum refund value of $0.02 for refillable bottles that are certified by the Oregon Litter Control Commission as being usable by more than one filler (the stubby, brown beer bottle). It is $0.05 for all other refillable bottles and all nonrefillable containers. Oregon also outlawed snap-top cans and at a more recent date banned aerosols containing fluorocarbon propellants from sale in the state.

Courtesy of Environmental Protection Agency

FIG. 6.1. SANITARY LANDFILL
Sanitary landfills offer communities the best currently practical methods of solid waste disposal. Use of new means of compacting and baling (shown) make landfill disposal more efficient. Properly managed landfill sites can be converted when filled, to parks and recreational areas.

TABLE 6.1. OUTLINE OF OREGON "BOTTLE BILL" (ORS 459.810 to 459.890)

The basic provisions of the Oregon law, commonly called the "Bottle Bill" are:
(1) "Beverage" is defined as beer, malt beverage, mineral water, soda water, and carbonated soft drinks.
(2) All "beverage" containers must have a minimum refund value clearly marked, paid by distributor to dealer, and by dealer to consumer.
(3) Dealers and distributors may not refuse to accept and refund the deposit on empty "beverage" containers of the kind, size, and brand which they sell.
(4) A "certified" beverage container is reusable by more than one manufacturer; capacity and shape may be set by the Oregon Liquor Control Commission.
(5) Minimum refund value of each container is 5¢, except for "certified" containers, for which the minimum refund is 2¢.
(6) Metal "beverage" containers with pull tabs that are detachable without the aid of a can opener may not be sold or offered for sale.
(7) Redemption centers may be established by any person in order to accept returned containers.

The Vermont law, which became effective on September 1, 1973, required a simple $0.05 across-the-board deposit on all beer and soft-drink containers. In 1975, Vermont passed new legislation which required that glass bottles must be refillable at least five times. It bans nondetachable ends and the plastic ring carrier for cans. This law became effective on January 1, 1977. (See Table 6.2 for European situation.)

TABLE 6.2. NON-RETURNABLE BOTTLES — EUROPE (1977)

Country	Status
West Germany	No regulations yet. Waste disposal act of 1972 allows government to ban special packages.
Belgium	1971 proposal to ban non-returnables not yet passed.
Denmark	Proposed to tax one-ways. A 1972 law allows government to ban one-ways — but not yet applied.
Finland	Proposal to tax non-returnables.
Norway	1974 laws non-returnables. Deposits on these packages are 20¢.
France	1975 law permits government to regulate or ban packaging based on ecological factors. No action yet. Tax expected.
Sweden	Since 1973, all packages for liquid products have been taxed up to 2 cents each. This tax might be increased.

It is important to note that both Oregon and Vermont are fairly low-population states with a deep concern for their environment. Vermont's added concern is about summer tourism and the need for clean roads and hiking trails. Passage of legislation similar to Oregon's and Vermont's for the more populated states, would have a devastating effect on the state's economy and on the economic health of the nation.

TABLE 6.3. APPROVAL/DISAPPROVAL OF BOTTLE BILL

		Approve	Disapprove	No Opinion
1973	All Oregon respondents	90.8%	4.7%	4.5%
1973	Ore. soft drink consumers	90.8%	4.6%	4.6%
1973	Ore. beer consumers	90.3%	5.0%	4.7%
1975	All Oregon respondents	89.9%	4.1%	6.0%
1975	All Washington respondents	68.0%	23.6%	8.4%

Courtesy: Applied Decisions Systems Study 1973 and Seattle Post-Intelligencer 1975

During 1976, both Maine and Michigan passed various laws calling for restrictive deposit legislation on beverage containers. These laws became effective in 1978. Michigan has also acted on the fluorocarbon problem in aerosols by requiring special labeling by January 1, 1978, and a ban on their sale by January 1, 1979, if potential harm to either health or the environment is established.

In 1973, Minnesota enacted into law a bill which was even broader, in nature than the laws of Oregon, Vermont, Michigan, Maine, Iowa, Connecticut and Delaware. It called for the *Minnesota Pollution Control Agency* (MPCA) to review all new and revised packages or containers except when such changes involve only color, size, shape, or printing. A new package, if deemed to be a problem, may be withdrawn from sale within 30 days after action by the agency. A ruling on a specific package may only be valid for two years which is until the new legislature meets in Minnesota. If a package manufacturer petitions for a package and the agency does not act for 120 days, the package then is immune for the next two years. At present, the legislature in Minnesota would like to review all packages sold in the state — old and new. The original 1973 law stated that any package or container sold at retail prior to May 25, 1973, was not covered.

As a means of evaluating the acceptability of a package, the MPCA formulated a check list called, "Criteria Evaluation Check-List for Packaging Materials and Substitutions." It contains various questions such as whether the package contains heavy metals, its litter potential, energy requirements, etc. Basic to the check-list is Question I which asks whether the social utility overrides all other conditions. If the package fails all the previous qualifications, it may still be sold in Minnesota if Question I is answered in the affirmative. In essence, Minnesota borrowed heavily from the EPA study and turned it into a state criteria for judging packaging. At this time, the Minnesota law is held up in court by a restraining order. No restrictive packaging legislation has yet been passed in Minnesota.

The present Carter Administration appears to be favorable to restrictive packaging legislation. It is anticipated that the next four years will see even greater restrictions on the free-flow of packaging.

Courtesy of Environmental Protection Agency

FIG. 6.2. MOUNT TRASHMORE
Mount Trashmore is an above ground sanitary landfill.

Weights and Measures

Although federal regulations are of vital concern to the packaging industry, many products are produced and consumed within *one state.* These items are subject to state legislation and thus cannot be regulated by federal regulatory agencies which apply themselves *only* to interstate traffic. Because of the great variance in the products, geography, and industrial productivity of all the fifty states, the laws relative to packaging and various products differ considerably on a state-wide basis. In order to reduce the degree of confusion in state and local regulations, the Association of Food and Drug Officials of the United States (AFDOUS) was formed. This group is composed of state officials primarily responsible for the enforcement of food and drug laws. The organization meets yearly with regional conferences set in alternate months. Specific requests for information should be addressed to:

Association of Food and Drug Officials
of the United States
Post Office Box 1494
Topeka, Kansas 66603

While state weights-and-measures officials are still primarily concerned with checking for short product weights, they have recently taken on a somewhat broader responsibility. Many have now started to handle other types of enforcement such as labeling regulations.

Recently, a group of 35 states went to the United States Supreme Court in an effort to render a decision as to whether states have the right to stop the sale of short weight products. The entire matter started in 1971 when the products of two companies were taken off the shelves in California due to short weight.

The two companies involved argued their case by pointing out that federal law allows them the right to sell short weight goods. They stated that variations due to loss or gain of moisture as well as unavoidable deviations in good manufacturing practice causes the short weight products.

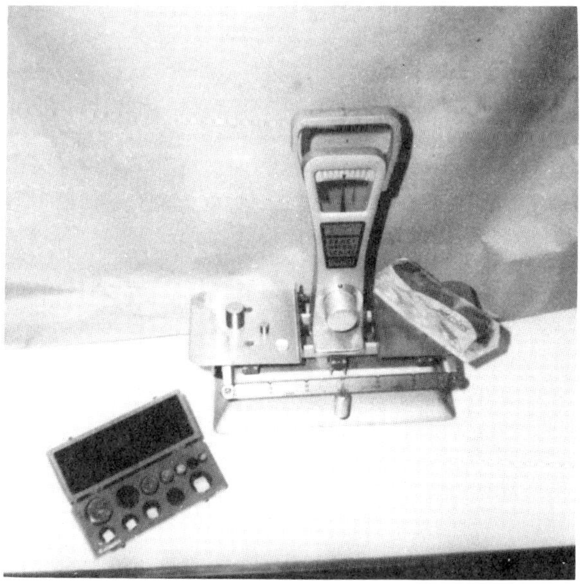

FIG. 6.3. GROSS WEIGHT DETERMINATION
First step in an official reweighing of a suspect commodity.

Courtesy of State of New Jersey, Division of Law and Public Safety

At issue is the inherent right of states to have the authority to check weights and order products off the shelf for inaccurate weight declarations. A decision in this case is due in the very near future.

In some instances, there have been overlapping responsibilities of both members of the *AFDOUS* and the *National Conference on Weights and Measures*. Also, they have been part of the same agencies. In general, both groups are independent with separate and distinct responsibilities.

As discussed above, the weights and measures laws of most states are not truly aimed at packaging but are concerned with the retail level. They are aimed at insuring that the consumer obtains the true weight upon purchase. States vary as to their weights and measures laws; California and Pennsylvania are fairly tough while New Jersey is "middle-of-the-road." A discussion of all states would be too unwieldy for this text; however, a sample state (New Jersey) will be treated.

Weights and measures in New Jersey are covered under the New Jersey Administrative Code (1364-4.7, 4.8, 4.9, 4.10, 4.11). As a subscriber to the model law, except where there is a difference between the model law and NFPLA, New Jersey is probably superior to other states in having qualified inspectors on duty. About 30 inspectors operate in New Jersey and check weights and measures on a retail level. At the invitation of a specific packaging plant, a state inspector may inspect and certify a plant scale. Adjustments can also be made and there is no charge to the plant. While this is a free state-wide service, a faulty scale does not constitute grounds for any court action.

FIG. 6.4. TESTING ACCURACY OF DELICATESSEN FAN-TYPE COMPUTING SCALE

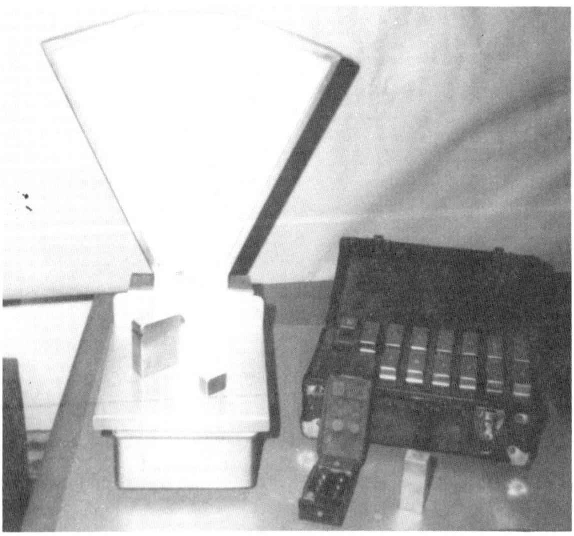

Courtesy of State of New Jersey, Division of Law and Public Safety

New Jersey also specifies certain containers for farm products. The wholesale containers for items such as apples, cranberries, mushrooms, asparagus, and corn are specified as to dimension and material.

Weight tolerance and deviation are stated as one-half of one percent of the printed weight on the package of every sample. Variations in manufacturing and distribution are also taken into account. This is a good system and not as rigid as Pennsylvania which averages 12 samples and requires that no single one fall in a tight differential.

Recently, the weight of school lunches was checked by New Jersey and prosecution started against school lunches deemed short in weight. This was successful and several suppliers were fined. An interesting point in New Jersey law states that in fresh meats, fat or any other material is to be considered as tare weight and not to be treated in the total weight. This is obviously impractical and not followed by state inspectors.

Additional clauses in New Jersey legislation requires that fresh crabs be designated per unit, shucked clams as liquid weight and if in shells, by unit. Pickles are noted as net weight.

Information about the weights and measures of New Jersey is available from:

New Jersey Office of Weights and Measures
Division of Consumer Affairs
Department of Public Safety
187 West Hanover Street
Trenton, New Jersey

Courtesy of State of New Jersey, Division of Law and Public Safety

FIG. 6.5. OFFICIAL LEGAL STANDARD BURETTES
Used by New Jersey State Office of Weights and Measures to officialy compare field flasks to the legal standards.

Other State Laws

The AFDOUS has been charged with the responsibility of working toward a uniform type law acceptable to all the states involved. While their goal is certainly an admirable one, it is currently far from reality.

About two-thirds of all states have passed a uniform law for food, drugs, and cosmetics. These laws have, in general, been based on the amended 1938 federal Food, Drug and Cosmetic Act. Other states have not adopted all of the provisions of model law. In states where the officials cannot adopt federal regulations, no provisions are made to prevent a manufacturer from using any additive. It is generally covered by a general phrase such as the use of any "non-poisonous material necessary to good manufacturing practice." Also, state laws reflect their special interests.

Some states have no laws, or very sketchy laws. In still other states, there is little or no enforcement. And inherent in most agencies, there is also a shortage of trained personnel capable of handling common day-to-day problems.

The ultimate aim is to allow state and federal officials to work in unison under basically the same laws and regulations. Some allowances as to proper jurisdiction would be done based on geography or type of industry. Full agreement on the interpretation of these laws and regulations would occur between all parties. The net result would be that every plant manufacturing confections would have some type of inspection by government officials. This would be true of states not only doing *interstate* business but also *intrastate* business.

In order to further examine current state regulations, several sample states will be discussed. While a comprehensive discussion of all fifty states is beyond the scope of this text, a look at several selected ones will illustrate the great differences involved.

In *Alaska,* labels showing ingredients are not required on baked products. Prohibited ingredients in confections include gypsum, barium sulfate, talc, chrome yellow. In addition, artificial flavoring, coloring and preservatives are required to be declared on all foods except butter, cheese, or ice cream.

California law states that confectionery products may contain artificial color, flavoring, and resinous glaze. *Massachusetts* also allows the use of "harmless" flavor, resinous glaze, refined petroleum jelly or refined mineral oil, natural wax, natural gum, and pectin in confections. Chewing gum may contain non-nutritive masticatory substances (actual types are not specified).

Nevada considers that confections containing alcohol or a non-nutritive substance, except harmless coloring, harmless flavoring, harmless resinous glaze, harmless natural gum, and pectin to be adulterated. Artificial color, flavor, or preservatives must be declared on the label.

Pennsylvania requires the use of "imitation" if an artificial flavor is used. Methyl salicylate, synthetic or derived from sweet birch, is not considered imitation. Coumarin is *specifically* prohibited.

Wisconsin set up standards in 1961 and these include cocoa, chocolate, and related products. Coumarin, extracted from tonka beans is permitted. This product is prohibited in federal law and specifically mentioned as illegal in many state laws.

Louisiana regulations state that a product is considered to be misbranded and/or adulterated if it is in violation of the federal law. Louisiana statutes cover the labeling of foods. The only additives mentioned are sodium benzoate, artificial coloring, artificial flavoring, and chemical preservatives, all of which must be declared. Also, their law states that packaged foods, drugs, cosmetics, and prophylactic devices be registered with the Board of Health, except stock feed. The latter must be registered with the State Department of Agriculture.

Local Regulation

In addition to states, many cities and counties have set up regulations that cover the jurisdiction of the broad range of consumer protection which includes packaging. These are often handled by the city's health departments which regulate food manufacture and packaging. The main concern of the city's officials are obviously those businesses that do not fall under either federal or state legislation. The enforcement of proper weights and measures is a basic concern of city officials. In the course of a year, many inspections are made of small to large stores, and scales checked for proper control. This may include retail outlets where bulk confections are sold.

The "Peek-a-Boo" law in New York City is an excellent example of a local regulation that affected packaging. This law states that all prepackaged meat must be displayed in transparent trays. Since its enactment in 1968-1969, many other localities have adopted this as law.

New York and Chicago have separate agencies that coordinate and oversee a wide range of consumer protection activities, including those pertaining to *deception* in packaging and labeling. These agencies do not get involved in matters of health or sanitation.

Many counties also have health departments. Los Angeles County has an agency which also has jurisdiction in the city of Los Angeles.

For further data on the above contact:

New York

Bureau of Food and Drugs
Department of Health
125 Worth Street
New York, New York 10013

Chicago

Chicago Department of Health
Civic Center
Chicago, Illinois 60602

Los Angeles County

Food and Drug Section
Bureau of Environmental Sanitation
Los Angeles County Health Department
220 N. Broadway
Los Angeles, California 90012

FEDERAL AND MILITARY

Totally different from other packaging forms, both federal and military specifications often impose severe problems on packaging suppliers. There are usually several steps necessary to be checked prior to fully understanding a federal or military order.

A question often asked by packaging engineers is why are military packaging regulations so complex and why do their packages consist of peculiar and little used materials. There are several factors that are unique to the military. Shelf-life requirements often are in excess of 10-20 years and the products used *must* be serviceable in a wide variety of climatic conditions. Army supply lines are often long and unyieldly, and may lengthen or change very rapidly. The logistics problem is often quite difficult to solve. The reliability of the packages employed can become as critical as the quality of the machines and the weapons they contain.

An additional factor is that the products and even more important, the military must be in a state of operational readiness at all times. Supplies must always get into the hands of the user in a serviceable condition. In conventional commercial packaging, the failure of a product to perform is largely a business loss or loss of good-will. For the military, it may mean the loss of an engagement and high casualties. The great variety of supplies involved coupled with the lack of knowledge of the likely distribution pattern for any particular supply means that the job has to be tackled in a different way.

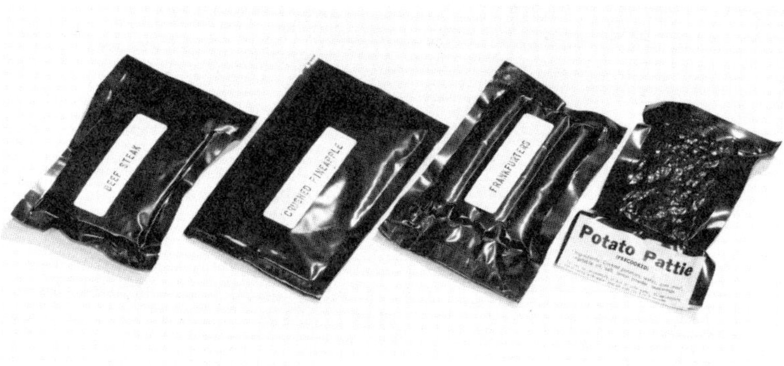

Courtesy of U.S. Army

FIG. 6.6. FLEXIBLE PACKAGES FOR THERMOPROCESSED AND DEHYDRATED FOODS

All federal/military items are listed by stock number in the "Federal Supply Catalogue, Stock List" (available from either the G.P.O. or U.S. Army Packaging School). Additional descriptions include whether the item is perishable or non-perishable, precise nomenclature, and overall packaging classification. In order to sell an item to the U.S. Government, it is mandatory to consult this list.

The list also refers to the four different types of specifications provided to packaging personnel interested in government affairs. Divided into *military* and *federal* specifications, each category has an *End Use Spec.* and *Packaging and Packing Spec.* (2 in 1 spec.). The End Use Spec. deals with the specific product while the Packaging and Packing Spec. refers to the packaging of the item and packing of the outer or shipping container.

Military specifications are identified by a MIL designation (or combination of letters) followed by either numbers or letters. Federal specifications are signified by a PPP designation followed by various letters. As examples, MIL-B-3877A refers to "Beef Slices and Potatoes with Gravy, Canned." PPP-B-636C refers to "Box, Fiberboard."

Also critical in understanding these regulations are the various levels of packaging imposed by the government. *Level A* is the most rigorous type of packaging demanded. It is intended for the protection of items to be shipped to various parts of the world and this may include both jungle as well as frigid conditions. *Level B* is generally less rigorous and denotes regular shipping conditions (RSC). The lowest type of packaging permissible is *level C.* This may include regular shipping containers and is the lowest type of package permissible for the best transportation rate. In general, the U.S. Government is attempting to phase out all level C type packaging. (See Table 6.4.)

TABLE 6.4. MILITARY PACKAGING STANDARDS

U.S. Level	U.K. Equivalent	Aim
A	Full Standard	Protection during worldwide distribution and storage anywhere in the world.
B	Limited Standard	U.S. distribution and during storage which is indoor (or covered) and is temperate.
C	Primary Standard	Temperate climate protection and regular shipping containers used (RSC).

In both MIL and PPP type specifications (and variations) the term "applicable documents" is found. This listing summarizes all the many federal and/or military regulations pertinent to the original specification. They all must be consulted prior to final governmental submission. Also necessary is the marking specification which specifies exactly how to mark the item for the various civil and military agencies. As a means of determining the rationale behind governmental packaging regulations, the U.S. Government has published the "Engineering Design Handbook

Courtesy of U.S. Army

FIG. 6.7. SHRINK WRAP TRAY

— Packaging and Pack Engineering." It provides the background for the criteria of military packaging and discusses items such as dessicants, corrosion inhibitors, and other factors peculiar to government and military packaging. It does not replace compliance with the proper documents, but does help understanding the overall problem.

In the U.K., the basic requirements for packaging intended for the military is found in *DEF-1234, General Requirements for Packaging Supplies for the Services.* It is a defense specification, issued under the sponsorship of the Military of Defense by joint agreement of all three defense services: Army, Navy, and Air Force. It is published in separate sections and consists partly of a Code of Practice. While all the data in DEF-1234 has complete Joint Service Agreement, it is sometimes necessary to supplement or even modify the instructions to meet local requirements.

In all federal/military packaging, one cardinal rule emerges. It is extremely important to check and verify both the packaging and marking of any item to be shipped with the proper government personnel. A mistake is very simple to make and the overall repercussions are severe.

Courtesy of U.S. Army

FIG. 6.8. TRIPLE WALL FIBREBOARD CONTAINERIZER, 40 × 48 × 43 in.

REFERENCES

ANON. 1975A. Legislative challenge; California weighs new anti-litter law. Mod. Brewery Age *26,* (47) Part 2, 32.

ANON. 1975B. How is the bottle battle going? Env. Sci. and Tech. *9,* (10) 906.

ANON. 1975C. The hang-ups on recycling. Env. Sci. and Tec. *9,* (12) 1015.

ANON. 1975D. New York labor, industry avail proposed returnable beverage law. Food & Drug Pkg. *32,* (10) 8.

ANON. 1976A. State scores: Mich., Me., yea; Colo., Ma., nay. Food & Drug Pkg. *35,* (10) 1, 10.

ANON. 1976B TRR banner yet waves over windy city. Canner/Packer *145,* (11) 22, 24, 26.

ANON. 1976C. State bottle bills: Taking the initiative. Env. Action *8,* (6) 8-12.

ANON. 1976D. A survey of consumer attitudes towards waste disposal and recycling. Aust. Pkg. *24,* (5) 36-37.

ANON. 1977. Mandatory deposits — The gameplan. *50,* (9) 16.

ATTMORE, M. 1976. Packagers find short weights costly in N.J. Food and Drug Pkg. *35,* (6) 1, 17, 27.

GILLIES, T. 1975. Pack groups file suit against MPCA. Food & Drug Pkg. *33,* (1) 1, 16.

HAMMOND, B. 1975. Recycling begins at home. New. Sci. *67,* (958) 152-153.

HICKOX, R. 1976. FEA study on container deposit looks at effect on energy, jobs. Food and Drug Pkg. *35,* (10) 1, 14.

JENNINGS, R. 1976. Bans on containers not an effective way to solve problems. Mod. Brewery Age. *27,* (48) Pt. 2, 36-37.

WAGGONER, D. 1976. The Oregon bottle bill—what it means to recycling. Comp. Sci. *17,* (4) 10-13.

7

Special Regulations

There are many different federal regulations pertaining to packaging which do not fit into any discrete category. Religious regulations form an additional separate and unique classification. These various regulations will be discussed in this chapter. Although some may appear to be of peripheral interest to the packaging industry, several are of direct value to users of packaging material.

CONTROLLED SUBSTANCES ACT — 1970

The *Comprehensive Drug Abuse Prevention and Control Act of 1970* was enacted into law as P.L. 91-513 on October 27, 1970. Title II of the law contains the *Controlled Substances Act (CSA).* Title III of the law contains the *Controlled Substances Import and Export Act.* Both titles of the law replaced seven major federal laws previously governing the manufacture and distribution of narcotics and other dangerous drugs. Certain provisions of the law became effective on October 27, 1970, while other provisions of the law became effective on May 1, 1971.

There are five classes of controlled substances as formulated by the CSA. Known as Schedules I-V, they are:

I Opiates, hallucinogens
II Opium derivitatives, coca
III Amphetamines
IV Barbiturates, chloral hydrate, paraldehydes
V Mixed narcotics

Exclusions include certain depressants and stimulants as well as any nonnarcotic substance which is classified as a non-perscription substance under the FDC Act. In the CSA, the Attorney General is given authority to classify narcotics and dangerous drugs according to their potential for abuse. The five above classes are the controlled substances.

In making his decision to control or remove a drug, the Attorney General must use the following criteria:

(1) Its actual or relative potential for abuse;
(2) Scientific evidence of its pharmacological effect, if known;
(3) The state of current scientific knowledge regarding the drug or other substance;
(4) Its history and current pattern of abuse;
(5) The scope, duration, and significance of abuse;
(6) What, if any, risk there is to the public health;
(7) Its psychic or physiological dependence liability;
(8) Whether the substance is an immediate precursor of a substance already controlled under this title.

The Attorney General also has authority to control a drug or substance which the United States is obligated to control under international treaties, conventions, or protocols.

A special symbol that conforms to 201K of the FDC Act must appear on the outside of each immediate container. Each label also must contain a symbol specific to the class of narcotic. Records must be kept and reports of every sale or delivery are required. No Schedule II substances may be distributed without a prescription, although Schedule III and IV substances may be so distributed if the FDA has determined that prescriptions are not required. The Attorney General is required to determine the total production amounts for substances under Schedules I and II and quotas are assigned which are limited to actual need. The quotas for each manufacturer are fixed before July 1 for each year for Class I and II narcotics.

The importance of the CSA and its relationship to everyday activities has recently been the subject of a court action. In mid-1977, a New Jersey man, under doctor's orders to take four "Librium" tablets a day, was arrested under a section of the Controlled Dangerous Substances Act of 1970. While traveling, the patient had transferred eight "Librium" tablets from his monthly supply of 120 tablets into an unmarked, one-in. green and brown vial. A neighborhood dog ran off with his denim jacket containing the vial. A woman who later found the jacket but could not learn the owner's name took it to the police, who found the pills and arrested the man. The arrest was made under a section of the CSA that prohibits carrying scheduled drugs in anything but the original labeled container provided by a pharmacist. The offense is an indictable misdemeanor carrying maximum penalties of three years in jail and a $1000 fine. The police contended that the original container provision is vital to the protection of children who might find a patient's unmarked pill container and unwittingly take damaging drugs. Without knowing the contents, doctors would not be able to treat them properly for dangerous or harmful reactions. The case was still in court in late 1977.

MILK AND TEA REGULATIONS

The *Filled Milk Act of 1923* prohibits milk in any form to which another fat or oil has been added. The *Federal Import Milk Act of 1927* provides for examination of cows by U.S. veterinarians or an authorized substitute. It establishes a bacterial count maximum for milk and cream and regulates the temperature of the product during shipment. Sanitary farm conditions are required. The *Tea Import Act* was enacted to prevent inferior quality in imported tea. Grading procedures were established for tea.

Other federal laws relating to food includes the *Poultry Products Inspection Act of 1957, Egg Products Inspection Act of 1970, Butter Act of 1923, Adulterated and Process or Renovated Butter (Internal Revenue Code of 1954)*, and the *Dry Milk Solids Act of 1944.*

FEDERAL CIGARETTE LABELING AND ADVERTISING ACT OF 1965

Designed to inform the public of a health hazard, this 1965 regulation established the use of a warning label on cigarette packages to read:

> "Warning: The Surgeon General has determined that cigarette smoking is dangerous to your health."

The lettering has to be in conspicuous and legible type on the package. Misleading labeling and advertising is specifically prohibited and cigarette advertising is prohibited on electronic media. There are also many other federal regulations dealing with record-keeping, tax stamps, marks, labels, and notices. These are found in Part 270, Title 27, CFR; Part 285, Title 27, CFR; and Part 270, Title 27, CFR (Export regulations).

FLAMMABLE FABRICS ACT

This act covers all wearing apparel or internal furnishing of natural or synthetic fibers, films or related paper, plastic, rubber, and foams. Under the CPSC, tests are developed and conducted to establish the criteria for dangerously flammable products. Many products are banned from sale because of their inherent flammability. Various labeling requirements are also established which must be used for these items.

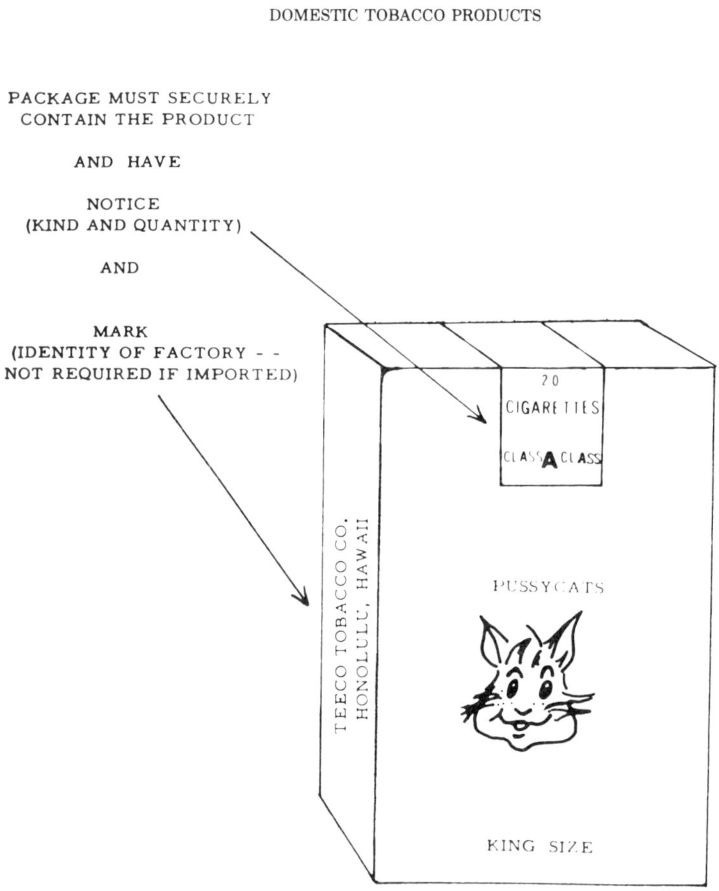

Courtesy of Dep. of Treasury

FIG. 7.1. DOMESTIC TOBACCO REGULATIONS
Federal regulations exist for the proper identification of tobacco products.

ALCOHOL BEVERAGE LAWS

The Bureau of Alcohol, Tobacco, and Firearms was established by Treasury Department Order No. 221, effective July 1, 1972. The order transferred the functions, powers, and duties arising under laws relating to alcohol, tobacco, firearms, and explosives from the Internal Revenue Service to the Bureau completely. On February 21, 1976, Treasury Department Order No. 221-3 (Revision 1) delineated to the Bureau those functions, powers, and duties relating only to criminal violations and forfeitures of the Federal wagering law provisions of the Internal Revenue Code of 1954.

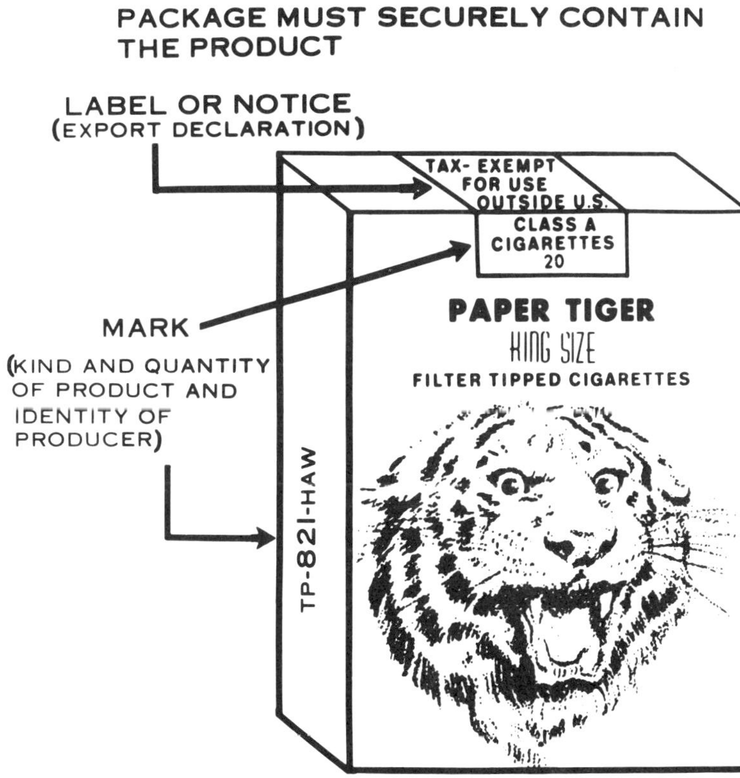

Courtesy of Dep. of Treasury

FIG. 7.2. EXPORT TOBACCO PRODUCTS PACKAGE
Export regulations for cigarette package.

Headquarters of the Bureau are located in Washington but since the Bureau is decentralized, most of its personnel are stationed throughout the country where many of its operational functions are performed. The Director, who is located in Washington, D.C., and who supervises the entire Bureau, is appointed by the Secretary of the Treasury and is under the supervision of the Assistant Secretary (Enforcement, Operations, and Tariff Affairs).

The Bureau is responsible for enforcing and administering firearms, explosives, and criminal violation and forfeiture aspects of the Federal wagering laws, as well as those covering alcoholic beverages and tobacco products.

The objective of the Bureau's program is to maximize voluntary compliance with these laws, and to minimize willful or involuntary violations of the laws. To achieve these goals, the Bureau is divided into two basic functions: criminal enforcement and regulatory enforcement. The objectives of the criminal enforcement activity are to eliminate illegal possession and use of firearms, destructive devices, and explosives; suppress the traffic in illicit distilled spirits; to enforce the criminal violation and forfeiture aspects of the federal wagering laws; and to cooperate with state and local law enforcement agencies to reduce crime and violence. The regulatory enforcement activity determines and assures full collection of revenue due from legal alcohol and tobacco industries; fulfills the Bureau's responsibility in the prevention of commercial bribery, consumer deception, and other improper trade practices in the distilled spirits industry; and assists other federal, state, and local governmental agencies in the resolution of problems relating to industrial development, ecology, and revenue protection.

Wine

Under the Federal Alcohol Administration Act (FAAA), standards of identity were established for wines. They are:

I Grape wines: table and dessert wines
II Sparkling grape wines, champagne and crackling wines
III Carbonated grape wines
IV Citrus and berry wines
V Fruit wines
VI Other agricultural wines
VII Aperitif wines
VIII Imitations
IX Retsina wines

The labeling requirements for wine are covered in Subpart D. Certain mandatory information is required and must appear on the wine label. This includes:

(1) Brand name
(2) Class and type
(3) Name and address of bottler and packer
(4) Alcoholic content
(5) Net contents in English and/or metric units
(6) Foreign wines must have name and address of importer

In Paragraph 4.39, an entire section appears detailing prohibited practices relative to the labeling of wine. No misleading brand names are allowed. No information can appear on the label that is in type smaller than 8-point Gothic caps — except on less than half-pint bottles. All required data must be in English. All statements that are "disparaging of a competitor's products" are specifically prohibited. Misleading containers are prohibited from use and fill tolerances are given for containers from 4.9 gallons to 4/5 quart.

Courtesy of the Council of Industrial Design (U.K.)

FIG. 7.3. ALCOHOLIC BEVERAGES
Wines, such as above, are covered by standards of identity.

Liquor

The labeling and advertising of distilled spirits is covered in Part 5, Subparts A-H of Title 27, Part 4. Subpart C details ten standards of identity for distilled spirits. They are:

I Neutral spirits or alcohol
II Bourbon and corn, blended and Canadian Whiskeys
III Gin
IV Brandy
V Applejack
VI Rum
VII Tequila
VIII Cordials
IX Flavored liquors
X Imitations

Courtesy of Reynolds Metals Company

FIG. 7.4. BOURBON WHISKEY
Labeling and overall product quality is strictly controlled by governmental regulations.

Subpart D deals with the labeling requirements of distilled spirits. Labeling is essentially similar to wine, except that the percent of neutral spirits and the state of distillation must be added. The label must remain intact except for vigorous washing with water and solvents. Gift cartons must bear mandatory information as must any opaque outer container other than a shipper. The standards of fill are similar to those of wine except for slightly tighter tolerances.

In early 1973, a North Carolina man died of acute ethanol poisoning after drinking half a fifth of whisky within hours after visiting his terminally ill wife in a North Carolina hospital. The man's family sought $700,000 in damages plus a requirement that distillers place warnings on the bottles that excessive and fast drinking of their product can be fatal. The suit was settled in mid-1977 in favor of the distilling industry. The U.S. Court of Appeals for the Fourth Circuit absolved the liquor company of any guilt in the death and ruled that distillers were not required to put health warnings on their products. This ruling is in contrast to the warning now required on cigarette packages.

Malt Beverages

Subpart C of Part 7, Title 27, deals with the labeling and advertising of malt beverages. They are quite similar to those of wine and liquor. The statement of alcohol content is required only by the state.

All alcoholic beverage labels must be approved and certified by the Federal Alcohol Administration.

RELIGIOUS REGULATION

Kosher Rules

Many Jewish consumers subscribing to Orthodox Judaism examine products for Kosher approval. A recent figure states that more than 2500 products from over 500 companies are manufactured according to the laws of "kashruth."

In the United States, about 45 large companies manufacture kosher-labeled processed meat products such as salami, weiners, and pastrami. The gross wholesale business of these firms averaged $70 million. The total retail volume of all kosher-labeled processed meats, including small plants operated by kosher butcher shops, is about $200 million. It is significant to note that a substantial portion of all kosher-labeled products are bought by non-Jews who have developed a taste for "kosher" meat products.

A kosher product must be packaged in an approved material which is subject to the same regulations as the product. Package approval is not easy to obtain but it is an absolute necessity for certain markets.

Definitions

Kosher. — A Hebrew adjective meaning "proper," used to describe a food or related product prepared and packaged in accordance with Jewish Dietary Law. Kosher is a religious term which has nothing to do with taste. The term kosher flavor is often confused with a garlic flavor. The two are not synonymous in any way, though some kosher foods may have a strongly spiced or garlicky flavor.

Kashruth. — The noun of Kosher. The five requirements of kashruth are the following:

(1) All pork products are inherently unable to be Kosher. The law also forbids consumption of birds of prey, e.g., owls, vultures, and eagles. Worms, snails, oysters, crabs, clams, and shrimp are also forbidden. The use of various byproducts of nonkosher animals are not permitted.

(2) Acceptable animals include the flesh of all quadrupeds with cloven hoofs, i.e., cattle, sheep, and goats. The barnyard fowl and scaly fish with fins are permissible.

(3) Fresh vegetables and fruits in their natural state are inherently kosher.

(4) The acceptable animals must be slaughtered and prepared according to certain regulations. If this is not followed, the entire food is nonkosher.

(5) No dairy and meat products may be mixed or consumed at the same time. Separate cooking vessels and dishes must be used for these products.

The above list is only a guide to the general requirements of the laws of kashruth. Other regulations exist.

The term "Glatt Kosher" is used to describe fresh meat from cattle or sheep whose lungs were checked by "bodkim" (trained religious Jews) and were found free of any adhesions. If the adhesions can be removed from the lobes of the lungs without leaving any trace of perforations, the meat is kosher. The term "Kosher-Style" on meat products is a euphemistic term for nonkosher and is very misleading. Its use is permitted in most states.

An additional term in kosher terminology is "parvae or parveh." This is applied to foods which can be classified either as dairy or meat dishes. They may be eaten with either type and do not infringe upon the dietary laws. Care is needed to keep parvae utensils separate and not to cleanse them together with either dairy or meat utensils.

The term "Kosher for Passover" means that there is no leaven in the food product. Due to religious regulations, it is strictly forbidden to eat leaven during the Passover season. Labels are attached to the package denoting Passover acceptability. Those are often pressure sensitive labels which can be removed after the Passover season.

The Interpretations of Kosher Laws

Since there is no authoritative hierarchy in the United States, an orthodox Rabbi's interpretation of the kosher laws is binding. It is important to note that not all Rabbis are qualified to judge the acceptability of a product. Judaism is divided into three basic sects: Orthodox, Conservative, and Reform. All Orthodox Rabbis subscribe to kosher laws and may interpret their tenets. In seeking approval, only Orthodox Rabbis should be consulted.

In the manufacture of food packaging materials and the packaging and processing of foods, a large number of chemicals are used. Each one must comply with kashruth. Cleaning materials must be free from any ingredients derived from animals forbidden by Jewish law. Emulsifiers, stabilizers, defoamers, and lubricants should be obtained from vegetable and/or synthetic sources. Glycerol widely used in cellophane manufacture, must be obtained from vegetables and not animals.

Although all natural fruits and vegetables are inherently kosher, processed fruits and vegetables are subject to investigation. Fresh, scaly fish is inherently acceptable, but not frozen fish. The preservative solution used for freezing must not have been used for freezing any nonkosher product. The freezing of poultry involves an analogous situation. Kosher poultry must not be processed with any chemicals or baths used for the nonkosher variety.

Aluminum foil must be rolled using synthetic oils. In some cases, high rolling temperatures may destroy animal lubricants and therefore the foil would be acceptable. Certain waxes derived from tallow are forbidden. No animal glues are acceptable. All plasticizers and flexible film additives are subject to investigation.

Acceptability is based on definition and interpretation of Rabbinical law. Many issues are decided after careful study and each product is a specific case. A layman cannot decide whether a product adheres to the Jewish law. Only a fully trained Orthodox Rabbi is capable of arriving at an answer. The steps involved are simple and the results may be significant. The difficulties in relating a modern product to ancient laws may be illustrated by an interesting example.

Most Orthodox Rabbis considered nonkosher any cheese made with rennet. Since rennet is derived from the stomach of a calf and non-kosher cattle are used, they reasoned, the product is not kosher. In order to process rennet, the stomach of a young calf is cleaned and stored for a considerable amount of time. Chemical additives are then added to yield a fine powder. Rennet is also used in pudding-type desserts. In biblical days, this process was unknown. A study made by a Rabbi determined that rennet is kosher — even if derived from nonkosher cattle. His reference was a quotation from ancient law stating, "The skin of the stomach is sometimes salted and dried so that it is rendered like wood, and if then it is filled with milk it is kosher, because after the stomach has been dried, it is just like wood, and does not contain any moisture of meat."

How to get Approval

There are three basic organizations offering converters a method of attaining kosher approval.

The Union of Orthodox Jewish Congregations of America created a national service in 1925. At the request of the interested company, a completed application detailing specific ingredients is submitted to the UOJCA central office. After an initial investigation of the product, the UOJCA sends rabbinical investigators into the plant. Supervision consists of observation of day-to-day practice and plant conditions. Packaging and processing equipment are carefully checked and investigated in order to determine compatibility with kosher laws. All data submitted and observed is fully confidential. If the laws of kashruth have been fulfilled, the company is permitted to use a (U) symbol on the specific products. the entire study involves a fee based on the total cost to the UOJCA.

The Joseph Jacobs Organization in New York City offers another supervisory service. An investigation and visit is made by an Orthodox Rabbi in order to determine adherence to kashruth. Approval is then given to display a "K" on the product. All information obtained is confidential and the fee is based on the total cost.

In Massachusetts, the Vaad Harabonium (Rabbinical Council) authorizes the use of a "VH" symbol. Certification is granted to products only if their manufacture remains under the strict supervision of the Vaad Harabonium. Companies are required to submit a detailed list of their ingredients and an investigation is made. Once a product has been cleared, the manufacturing plant and equipment are cleaned and sterilized. Manufacture then starts under the supervision of an Orthodox Rabbi assigned to the plant's production and quality control departments. As in the prior cases, all information is confidential and a rated fee is paid.

In addition to the above three organizations, an agreement may be made with an individual Orthodox Rabbi for approval. The degree of supervision is wholly dependent on the Rabbi. Approval will allow for his signature on the package vouching for its kosher purity.

When a product has been granted approval for adherence to the kosher laws, the three organizations provide a variety of marketing and counseling services for the product. The potential increase in sales from obtaining approval may be substantial.

Other Rules

In contrast to the rules of Orthodox Judaism, no other religion has formulated a set of rules for food packaging in the United States. (See Table 7.1.) This may be due to the lack of immigrants from other than Chris-

TABLE 7.1. ENDORSED CONSUMER PACKAGING MATERIALS[1]

Aluminum Foil
Plastic Wrap & Bags
Diamond Foil
(Allied Aluminum
Fabricators Co.
Louisville, Ky.)
Amax
Food Service Wrap
(Amax Aluminum Co.,
Food Service Div.
St. Louis, Mo.)
Reveal Roasting Wrap
STRETCH 'N SEAL
Baggies (Polyethylene Bags
Sandwich
Food Wrap
All purpose
(Colgate-Palmolive Co.
N.Y.)
Saran Wrap
Handi-Wrap
Sara-Stretch
(Dow Chemical Co.
Midland, Mich.)

Mobil
Plasticware
(Mobil Chem. Co.
Macedon, N.Y.)
Presto
Bags
Plastic Wrap
(Presto Prod.
Appleton, Wis.)
Reynolds Wrap
Food Service Film
Interfolded Sheets
Interleaved Sheets
Potato Wrap
Vented Foil
Brown-In-Bag
Oven Bags
Reynolds Consumer
Packed Pans
Broiling Pans
Cake Pans
Loaf Pans
Pie Pans
(Reynolds Metals Co.
Richmond, Va.)

Job Squad
Paper Towel
(Scott Paper Co., Inc.
Phila., Pa.)
Topco
Bags
Plastic Wrap
(Topco Asso.
Skokie, Ill.)
Glad Plastic Bags
All Purpose
Food
Sandwich
Glad Oven Bags
Wrap
Kitchen, Freezer & Refrigerator
(Union Carbide Corp.
N.Y.C.)

[1] From "Kosher Products and Services Directory, Consumer Edition (1974-5)"

tian lands. Until recently members of the Roman Catholic faith could not consume meat on Fridays, but this did not prohibit the use of packaging materials containing glycerine from animals. Kosher meat products are bought by Seventh-day Adventists and Seventh-day Baptists who observe the Sabbath day on Saturday. They do not eat any products which may contain an ingredient of pork or lard. Thus religious precepts differ not only among faiths but also in the degree of strictness of adherence to the precepts.

Although not formalized in the United States, members of the Hindu and Islamic faith follow strict rules as to food and indirectly to packaging.

REFERENCES

ANON. 1975A. Packaging laws chart. Mod. Pkg. *48*, (12), 18.

ANON. 1975B. Washington survival guide. Pkg. Eng. *20*, (5), 36-39.

ANON. 1976A. Liquor containers — in the metric spirit. Mod. Pkg. *49*, (4), 5.

ANON. 1976B. Wine case marking must now reflect metric contents. Food and Drug Pkg. *35*, (2), 3, 10.

ANON. 1976C. FDA Proposal to remedy drug-packaging ills. Mod. Pkg. *49*, (4), 12, 52.

ATTMORE, M. 1976. FDA regulations require expiration dates for drugs. Food & Drug Pkg. *34*, (6), 1.

DAVIS, R.D. 1975. The challenge of ingredient labeling. The Brewers Dig. *50*, (6), 28, 34.

HECHT, H. 1976. Informing patients on prescription drugs. FDA Consumer. *10*, (2), 4-7.

HICKOX, R. 1976. HEW seeks to ease rule making. Food & Drug Pkg. *35*, (6), 1, 20.

HINKLE, R.P. 1976. Wine labels: What they don't mean. Wine Word *5* (3), 32-33.

MARDER, H. 1976. Laws/Rulings: Just another bureaucratic onion. Mod. Pkg. *49*, (11), 8.

MITTELBERGER, E.G. 1976. A bicentennial of wine packaging. Good Pkg. *37*, (6), 10-11.

SACHAROW, S., GRIFFIN, JR., R.C. Food Packaging, Avi Pub. Co., 1970.

SMELING, H.V. 1976. FDA enforcement activities defined at food and drug lay institute session, Food Proc. *37*, (5), 10-11.

STEVAN, M.A. 1974. What shoppers are concerned about, FDA Consumer *8*, (7), 13-17

8

Carrier Rules

Packaging regulations are not solely concerned with the final consumer package. The method by which this unit reaches the marketplace often plays a vital part in its safe arrival. There are regulations that determine both the cost of transportation and outer shipper construction for most packages. Of greatest importance to the packaging engineer are *rail* and *truck* regulations. These will be discussed as well as the regulations covering water and air transportation.

Shipping regulations and the establishment of tariffs for passenger and freight rates for land and water transportation are covered in Title 49, Transportation, Parts 1300-end. All companies involved in transportation must publish scheduled rates for their services. Acting as common carriers, these charges describe the services, their limitations, conditions, and establish freight rates.

Products are classified into discrete categories based on value, fragility, density, potential for damaging other cargo or the carrier's equipment, and the frequency and direction of shipping. These separate categories have one or more rules which designate the type of package to be used. A category is assigned a rating or percentage fee of the first class or standard freight rate.

It is essential that these rules be followed since failure can result in penalty charges, refusal of shipment, or refusal of a freight damage claim. There are seven rules governing packaging in the railroad regulations and thirteen in the trucking regulations.

RAILROAD REGULATIONS

The minimum standards for corrugated boxes are found in Rule 41 of the *Uniform Freight Classification.* This committee has been set up by the railroads (Association of American Railroads) to protect them from severe damage claims resulting from improperly used corrugated shippers. The group also issues tariffs for the railroads, subject to Interstate Commerce Commission (ICC) approval.

In rail shipment, three main hazards exist for packages. When rail cars are assembled in yards, the shunting operations may produce impacts with other goods. Snatching also occurs which results from starting and stopping loosely coupled wagons. Vibration depends on the speed, type of car, and condition of the track. It is probably the greatest hazard in rail shipments.

Courtesy of Association of American Railroads

FIG. 8.1. MODERN RAILROAD PIGGYBACK TERMINAL

Rule 41 defines the minimum acceptable standards to be used for rail shipments. It is essential that the corrugated container used not merely meet Rule 41 regulations. The shipper must also comply with the conditions imposed by the specific product. Presently, nearly all corrugated shippers are made to meet Rule 41. If Rule 41 is not met, there are scheduled upcharges for freight tariffs.

In Rule 41, the actual specifications for acceptable corrugated containers are published. Included are specifications for packing fragile items, cushioning and inner wrappings, types of boxes and their details, methods of closure, joint construction, and liquid packaging. Also noted in Rule 41 is a method of drop testing. Under Rule 41, the drop test states:

Under 20 lb — 16 in. drop
20 to 30 lb — 14 in. drop
Over 30 lb — 12 in. drop

Ten drops are conducted on both top and bottom with eight flap edges.

Shippers are determined by noting the size of box — its sum of length, width, and depth. Needed is information on the gross weight of the box and its product. By then checking Rule 41 tables, minimum combined weight of facings, burst strength, and minimum puncture of triple wall are found. When used, a certificate seal notes adherence to Rule 41 compliance. If the product weight exceeds the table, an alternate package may still be used among special numbered containers.

Included in the *Uniform Freight Classification* are additional rules of interest to the packaging engineer.

Rule 6 requires that fragile items or items in glass or earthenware be marked fragile or with a similar marking.

Rule 21 defines "nested" articles.

Rule 22 designates various types of finishes on wooden articles, i.e., furniture.

Rule 40 covers the entire area of shipping containers and notes the specifications for pails, crates, barrels, drums, canboys, corrugated boxes, and multiwall bags.

Rule 49 covers test or experimental shipments. It is quite useful for products not covered by standard procedures. Stated in the rule is the need to petition the chairman of the *Uniform Classification Committee* as to approval for shipment. The data must include a description of the container, product, and details about loading and bracing. The petition must be submitted two weeks in advance to obtain approval. The purpose of Rule 49 is to relieve the carrier of liability with respect to failures in container, loading, or bracing procedures.

The packaging of hazardous materials for rail shipment is described in the Rail Carrier Tariff and Water Carrier Tariff. This is probably the best place to find all the data necessary for the proper transport of hazardous materials by rail. Also available is the Uniform Freight classification organized by the railroads to protect them against damage claims due to improper shipping containers (See pg. 135). Contained in this specification are both tariffs and actual package construction data for hazardous materials.

In early 1978, three rail accidents occurred involving hazardous chemicals. Materials included liquified propane, sodium hydroxide, and chlorine gas. Although any chemically dangerous substance can be packaged and shipped safely and routinely, if the integrity of the container system is destroyed by derailment, trouble can occur. It is expected that these accidents will lead to new and revised standards for the handling of hazardous materials by rail. (See Table 8.1.)

TABLE 8.1. MAJOR ACCIDENTS WITH HAZARDOUS MATERIALS

Date	Material	Transport Used
April 16, 1947	Ammonium Nitrate	Water
February 24, 1978	Liquified Propane	Rail
February 26, 1978	Sodium Hydroxide	Rail
February 26, 1978	Chlorine Gas	Rail

Note: Total Rail Derailments (1977): 7,858
Rail Derailments Involving Hazardous Materials: 500

In the U.K., the transport of hazardous materials by rail is governed by the British Rail's conditions of acceptance. Described in "Dangerous Goods by Freight Train and by Passenger Train," the individual products are classified by hazard and the packages which will be accepted are specified. The pictorial symbols recommended for international shipment by the United Nations have been adopted and combined in a single label with the railway's own loading instruction labels.

In an actual case, the "Shrinkase" program[1] (involving the use of shrink film with paperboard) was strongly protested by the Paper Box Manufacturers' Association. If the quantity of paper board used in the two trays of the "Shrinkase" package versus the quantity used in the conventional paperboard box is compared then it is readily seen that less paperboard is used for "Shrinkase." Despite this opposition, classification approval was obtained on all canned vegetables and it is shown as Package 793 in the Uniform Classification. It also appears as Package 500 in the National Motor Freight classification. (See pg. 137.)

[1]RTM of Reynolds Metals Company

Courtesy of Association of American Railroads

FIG. 8.2. SMOOTHER RIDE
By welding sections of rail together in continuous lengths, railroads reduce track maintenance costs and give their customers a smoother ride.

MOTOR FREIGHT REGULATIONS

The National Motor Freight Traffic Association, Inc., publishes the *National Motor Freight Classification* through a committee organized to prepare this list of tariffs. The various tariffs are then filed with the DOT for freight utilization and ICC approval.

For New England truck carriers and for shipments originating in or destined for New England areas, the tariff used is the *Coordinated Freight Classification.*

All tariffs for motor freight transport are set up by the American Trucking Association through the above mentioned committees.

Similar to Rule 41 for rail shipments is Rule 222 for the trucking industry. This rule contains the essential information found in Rule 41 except for several minor variations.

The main hazards in road transport are vibration and bouncing of the load. Bouncing due to large scale transportation in the road surface is not a major problem. Load crushing by ropes, shifting of the load, and impacts from other goods are quite important.

After a new package has been perfected and ready for marketing, it must be presented to the National Classification Board for their approval as to truck shipment. This group, part of the National Motor Freight Traffic Association, Inc., may issue a test permit for shipping purposes in order to determine the durability of the package in actual transportation. Regardless of how good a package is developed, it cannot be shipped legally in interstate commerce via common carrier without proper classification approval and the classification tariffs properly referenced.

In order to get classification approval, a test permit is issued to the party that will actually ship the commodities packaged. On each shipment made, the bill of lading must be referenced with the test permit number and inspections made at destination and reports prepared on each shipment for submission to the classification committees.

After a sufficient number of shipments have been made, application is made to the proper classification committee and supported with the test

Courtesy of American Trucking Association

FIG. 8.3. ROCKING HORSE
A rocking horse is properly crated for freight shipment.

reports, requesting classification approval of a new package. If the committees are satisfied with the reports and feel that a sufficient number of test shipments have been made, they will then place the new package, as specified, on their docket for hearing. At the hearing, further evidence must be presented in favor of the package — opposition is usually expected, even on a new innovation in packaging.

It is important to note that the above description of the method necessary for obtaining approval is valid for both *rail* and *truck* shipment. The tariffs noted by the classification committee are not the same as the hazardous articles tariffs under the DOT. Both sets should be studied prior to making a decision.

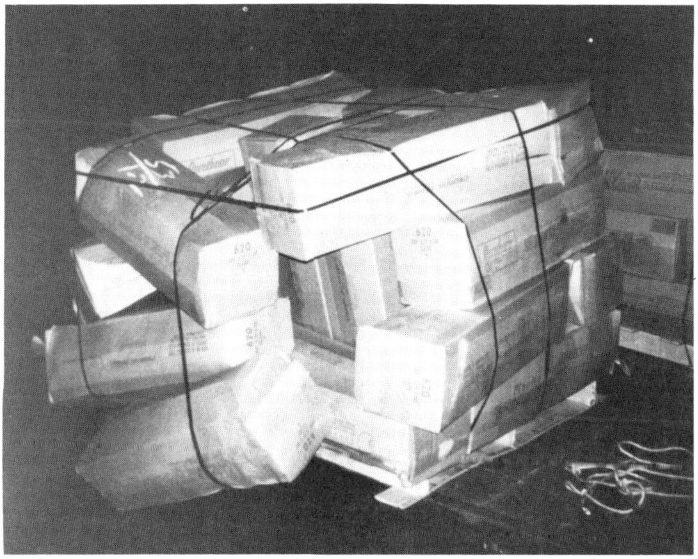

Courtesy of American Trucking Association

FIG. 8.4. POOR ARRIVAL
This pallet has been poorly strapped for truck shipment. Note its unsatisfactory arrival.

WATER

Shipment by sea is largely international in scope. The International Maritime Consultatives Organization (IMCO) details the shipment of hazardous materials by sea. In the U.S., the Water Carrier Tariff has now become the standard for shipping all hazardous materials.

AIR

On passenger planes, there are more restrictions than on cargo planes. No detailed package specifications exist for most products and the airlines will accept any package deemed satisfactory for the purpose used. Almost all explosives and all dangerous poison gases are prohibited from air transport.

The specific air transport rules are governed by the Air Transport Association (ATA) in the United States. For foreign transport, these regulations are administered by the International Air Transport Association (IATA). Both groups publish suggested specifications only for live animal shipments and hazardous materials. Also available is "Packaging for Profit" from the ATA. It is a guide toward suggested packaging procedures.

UNITED PARCEL SERVICE

For domestic shipments, the United Parcel Service (UPS) transports packages in a very efficient manner. Packages with gross weights of over 50 lb are not accepted and their size limit is 100 united in. and no more than 100 lb in a single shipment. United in. equals length plus girth, where length is longest side and girth is the circumference at the thickest part.

In addition, the UPS maintains extensive facilities for testing and publishes some very useful books on packaging and shipping.

PARCEL POST

Parcel post regulations are covered in the U.S. Postal Service Manual, Sections 121.3-122.16. The maximum weight for postal shipment is 40 lb and 84 united in. Parcel post is used for small individual packages over 1 lb. Second through fourth class weight limitations are 70 lb and 100 united in.

For hazardous materials, the basic regulation is in Title 18, Chapter 83, Section 1716. It is quite restrictive and generally bars most hazardous materials from mail transport.

REFERENCES

ANON. 1975. Calculating your liability risk. Mod. Pkg. *48,* (10) 10-14.

ANON. 1976A. A guide to government packaging regulations: Hazardous materials and carrier rules. Part 6. Package Printing and Diecutting *22,* (9) 8-9,57.

ANON. 1976B. The transport of merchandise in containers. Emballage *46,* (335) 39-43. (In French).

ANON. 1976C. Fibre box association proposes study for changes in Rule 41 and Item 222. Paperboard Pkg. *61,* (6) 20, 22, 24, 27.

ANON. 1976D. Design for printing a shipping symbol requires knowledge of box-making practices. Paperboard Pkg. *61,* (4) 50, 52, 54, 57.

ANON. 1976E. Shipper test standards near. Mod. Pkg. *49,* (2) 16-17.

DIXON, J. 1975. Packaging for profit. Dist. World *74,* (11) 23-28.

FAHY, E.R. 1976. Building up on many changes in hazardous materials regulations. Mod. Pkg. *49,* (9) 27-28.

GILL, L.S. 1976. Beware of booby traps in multinational distribution. Hand. and Ship. *17,* (3) 44-45.

LAMB, R.C.S. 1976. Planning and management in merchandising. Aust. Pkg. *24,* (7) 35, 37.

MALTENFORT, G. CO., 1976. A proposal for the modernization of United States carrier rules for corrugated shipping containers. Pkg. Dev. and Systems *6,* (5) 32-37.

MALTENFORT, G.M. 1977. Analysis of overseas carrier rules show many differences from U.S. practice. Part 2, Carrier rules in Great Britain. Paperboard Pkg. *62,* (2) 52-55.

MCMASTER, R. 1976. Reusable shipping containers. Meat Ind. *22,* (8) 28.

MOORE, W. 1976. Protective packaging impacts on PDM. Hand. and Ship. *17,* (10) 49-56.

SIMONDS-SHEA, R.A. 1976. Packaging fragile goods for export. Pkg. *47,* (554) 21-22.

SOLAK, W. 1976. Motor carrier representative says much improvement is needed in vehicle suspensions for shipping safety. Paperboard Pkg. *61,* (11) 36, 38, 43, 47-48.

9

Patents, Trademarks and Copyrights

PATENTS

The concept of patents is not new. As early as the 14th Century, British monarchs granted exclusive privileges to stimulate commerce and industry. Patents were granted in colonial America as early as 1641. In 1790, the first U.S. patent law was passed. After major changes in 1793 and 1836, the patent laws remained basically unchanged until 1952. At that time they underwent great changes. Originally, the Patent Office was located in the Department of State. In 1849, it was in the Department of Interior and finally, in 1925, it again changed to its present home — the Department of Commerce.

A patent is a grant by a specific government of certain exclusive rights to an inventor in exchange for making his invention public. In the U.S., a patent is granted for 17 years. During this period, the inventor holding the patent can prevent others from making, using or selling his invention without his consent. After this period, the invention can be used by anyone. (See Table 9.1.)

While there have been over four million patents issued in the U.S., many have not seen the "light of day." Most are non-commercial and totally impractical. About 700,000 patents are still in force — about 50% are being practiced.

TABLE 9.1. TYPES OF INVENTIONS

Type	Example	Comment
Secret	Coca-Cola formula	Difficult to maintain
Public	Polio vaccine — Dr. Salk	Noble
Patent	Polaroid Camera — Dr. Land	Most Popular

Patent Consideration

To be considered for patenting, an invention must be new and *useful* and not an obvious way of doing something in order to be patented. Ideas or suggestions, business methods, printed matter or simple formulations cannot be patented. No patent can be issued if the innovation is known or published in the U.S., or abroad more than a year before patent application in the U.S. including any disclosure by the patentee. The invention must be a true advance over prior art and not simply a minor advance which would be obvious to a person with ordinary skill in the art. Changes in size or materials of construction are rarely patentable.

Several categories are used for patent consideration. It must fall into a manufactured product, composition of matter, process, machine, plant, or design classification. A new package made from a plastic is a manufactured product. The plastic alone comes under composition of matter. The combination of steps used in manufacturing the package is the process. The equipment for producing the package is in the machine category. A new variety of rose is in the plant category while a new package design is in the design category.

Courtesy of U.S. Patent Office

FIG. 9.1. PATENT AND TRADEMARK OFFICE
An agency of the U.S. Department of Commerce, this office is located at 2021 Jefferson Davis Highway, Arlington, Virginia.

Information on patents are obtainable from the *Official Gazette* of the U.S. Patent and Trademark Office. This is a weekly publication that abstracts granted patents with selected figures. Also included is information on patent notices and suits and changes in the rules and regulations of patent issuance.

Patent Procedures

The first step in obtaining a patent is to establish the actual time of conception. This is often called the disclosure of invention and establishes the time of conception of the specific invention. It must be signed and dated in the presence of a witness.

A patent search then follows. Usually, a preliminary search is conducted to determine whether the invention has already been patented. A group of similar type patents may be found and can help determine whether a full scale extensive search should be made. Also useful will be any data that can help the inventor write the patent in a different manner less open to possible contest. A full search of the patent literature is made to determine whether the invention has already been patented.

There are many public facilities where searches are conducted. The search room in Arlington, Va., is a major facility and about 25 university or public libraries offer extensive patent files. To facilitate both a patent search and patent preparation, a special patent attorney or patent agent is often used. These people are licensed by the U.S. Patent Office. A directory of licensed personnel is available from the Patent Office.

The actual patent consists of two basic parts. The specification is used to allow the reader to understand the invention and utilize the concept. The claim details what is actually covered in the patent and is often written as broad as possible.

The application is a written document, signed by the inventor or joint inventors, and contains a full description of the specifications and claims for the invention. It should state the difference between this patent and all other similar inventions. At one time, models were required; however, they are no longer demanded by the Patent Office. All feasible drawings should also be included in the application. The filing fee is $65 plus $2 for every claim over ten. Finally, the inventor's signature is necessary prior to submission.

After arriving at the Patent Office, the application is put in arrival order and usually examined in about 3 months by a patent examiner. He searches the patent files and notes whether any claims made are already in the literature. Usually certain claims are rejected and the examiner's decision follows shortly. Rejection is not unusual in the first action. The applicant can then reply to an adverse action and request re-examination with or without amendment to petition. It is essential to respond to every objection in the first action. This procedure usually occurs in 30 days to 6 months but is commonly done at a 3 month interval. The Patent Office's second action is usually final.

There is also provision for appeal. The Board of Appeals may be utilized and action can be brought against the Patent Office in the U.S. District Court of Washington, D.C.

Even after a patent has been issued, problems may occur for the inventor. Interference can occur within one year of a published and accepted patent. A determination is made of who came first in arriving at two similar applications. In this case, the disclosure of invention is critical. Notes, drawings, meetings are all essential in determining a date of conception and reduction to practice.

Most patents are issued to the inventor but assigned to his employer. Since a patent is considered to be real property, it is also capable of sale, mortgaging, and bequesting. Certain parts of a patent may be assigned or specified to be used in exclusive areas within the U.S.

Patent infringements can be sued by either party in a U.S. District Court with appropriate appeals in federal Courts of Appeal. The U.S. Patent Office does not take part in such suits.

Courtesy of U.S. Patent Office

FIG. 9.2. CONFERENCE
An examiner discussing the merits of an application with an attorney.

A patented article must be marked with the patent and number. The terms "patent applied for" or "patent pending" have no legal significance. It merely means that action has been taken in applying for a patent and it may be an inhibiting factor towards someone stealing the idea or concept.

Also issued are design patents which cover an original, ornamental design for an article of manufacture. It does not cover the structure or utility of the design.

On an international basis, foreign protection requires filing in each country. The laws vary widely between nations. Most countries are signees of the *International Convention for the Protection of Industrial Property,* Article 4. These nations require the invention to be produced in the country within 3 years. The Paris Convention or Agreement was signed by 79 nations.

In 1970, a New Patent Cooperation Treaty was signed by 35 nations. When the treaty becomes effective, it will centralize filing and standardize the format of application.

The inventor must have a U.S. license before applying for a foreign patent if filing before the U.S. application or within 6 months of U.S. application.

An additional term of considerable interest to patent holders is the matter of licensing. A license is the granting of permission by the patent

Courtesy of U.S. Patent Office

FIG. 9.3. SEARCH
The Patent Office's Public Search Room, contains in classified subject matter, category patents on every conceivable device known to man. More than 12 million U.S. and 7 million foreign patents are on file and available to the general public.

holder to use his invention. If the invention had been used without permission, infringement would have occurred. This permission is normally granted for the payment of some type of fee or royalty.

Licenses are a most important part of common industrial life. They are often beneficial to both parties involved. The person granted the license is given great freedom to operate while the patent holder derives an income from the sale of the license. Often the holder is not ready nor does he want to operate his license. He still holds absolute control over his invention but is making his invention pay off profitably.

TRADEMARKS

Modern packaging is based on the concept of trademarks. Without a trademark, there would be no incentive for manufacturers to produce unique packages for their products. The product is packaged and then sold in order that it evoke a pleasant memory in the consumer's mind. *It is absolutely essential that there be legal protection and/or registered trademarks available for all manufacturers.* The development of a trademark often costs a great deal of money, but sales can be substantially increased by the use of a suitable trademark. (See Table 9.2.)

The Trademark Act of 1905 authorized the registration of trademarks in interstate commerce. Altered by the Lanham Act of 1947, it called for provisions for the registration of certain service marks and cancellation of registration marks.

A trademark costs $35 and is in effect for 20 years. Renewals are permitted for additional 20 year periods. The mark is then published in the *Official Gazette.*

In the United Kingdom, the registration of trademarks is covered by the Trademarks Act of 1938. A specific registry, Trademarks Registry, administers this Act. Registration is essentially similar to the U.S. Although the initial term of registration is seven years, the registration can be renewed indefinitely for further periods each of fourteen years.

What constitutes a trademark is also of importance. Any name, word, symbol or combination that is used by a manufacturer to identify that his product is different from another, may be defined as a trademark. Difficult to protect trademarks would be geographical areas, people's names, etc. Use of "T.M." when the name is employed is considered to be beneficial in strengthening possible future claims. It is not necessary to use "T.M." everytime the name is written — but it should be used enough to show that the manufacturer does have a certain right to the name.

Many trademarks have become accepted names. Cellophane is used generically in the U.S.; however, in the U.K., the word "cellophane" is a registered trademark of British Cellophane, Ltd. The term regenerated cellulose is used as the generic name in the U.K. The word Saran was coined by Dow Chemical to describe their copolymer of vinyidene chloride (PVDC). In order to avoid its becoming a generic term (which it is), it was later changed to Saran Wrap. Nylon is the common term for polyamide films. It also has become a commonly used generic term. It is often useful to use the name *brand* after the trademark to show originality. An excellent tensile tester is made by the Instron Co. Its use should be specified as an "Instron" tensile tester since other firms make tensile testers.

TABLE 9.2. HINTS FOR SUCCESSFUL TRADEMARK USE

(1) Give trademarks distinctive graphic treatment
(2) Use a trademark as a capitalized proper adjective
(3) Be consistent in use
(4) Never use a trademark as a verb
(5) Use a correct generic name with the trademark — if possible and practical
(6) Accompany all trademarks which are registered with registration notices
(7) Make sure all firm employees protect the trademark
(8) Never use a trademark to describe something for which it is not intended
(9) Alert the legal department whenever you notice an improper use of the trademark
(10) File Federal registration during the sixth year after registration every 20 years
(11) If a trademark is modified, a new registration may be needed
(12) Keep up with all changes in trademark law by requesting annual review by legal personnel

COPYRIGHTS

The landmark legislation in copyright activity was based on the Universal Copyright Convention held in Geneva in 1952. Revised in 1971, all the new material became effective July, 1974.

The basic objective of the convention was to integrate existing national systems of copyright on the basis of a simplified reciprocal national treatment.

By law, the copyright owner is entitled to:

(1) Print, reprint, publish, copy, or sell his copyrighted work
(2) To translate, to convert into a novel
(3) To deliver or authorize the delivery of the work in *public for profit*

Other provisions exist but it is essential to note that the concept of copyright gives the owner a commercial interest in the specific act involved.

Copyright lasts for a 28 year term from date of first publication. An extension of another 28 years is possible upon the submission of proper applications. Pending the introduction of a new copyright law, the copyright of works whose second term of 28 years would have expired is being extended automatically so that they may enjoy the new term of copyright when it is introduced.

Copyright may be "assigned, granted or mortgaged by an instrument in writing signed by the proprietor of the copyright."

Information cannot be copyrighted. It is only the specific form in which it is presented that can be subject to copyright law. Also, while lectures may be copyrighted prior to delivery, when published at a later date, they must be copyrighted again. In order to be copyrighted, material must use the word "copyright" or © near the name. This shows that the material has been copyrighted.

In the packaging industry, the effect of copyright is rather negligible. As discussed before, specific photos, illustrations, puzzles, etc., appearing on a package might be subject to copyright. Court cases have come up when a photo was used on a package and it was the copyrighted property of another owner. These problems can and often do result in considerable expense to all parties involved.

An interesting case came up in the U.K., which does illustrate the concept of copyright in packaging. In the case of *Kenwick vs. Laurence,* cards were involved which showed a hand holding a pencil in the act of completing a cross in a square. These cards were intended to show illiterates at election time how to vote. Developed by Kenwick, Laurence published a similar card with a similar illustration, but the hand was in a somewhat different position. The courts ruled that while the idea was the same, the second card *did not* infringe on the first because an idea cannot be copyrighted. The cards were not identical.

It is also quite important to obtain permission for the use of any testimonial letter on a package. Although the writer may be quite flattering, the owner of the letter is the writer and permission *must* be obtained prior to use. This also holds true for picture reproduction on a package.

Additional Reading

For both background information and more detailed data, the reader should consult the following references:

PATENTS

(1) *The Official Gazette.* A weekly publication detailing each patent granted that week. Published by the U.S. Patent Office, Washington, D.C.

(2) *General Information Concerning Patents.* Available from the Supt. of Documents, G.P.O., Washington D.C. 20403 or the nearest field office of the U.S. Dept. of Commerce.

TRADEMARKS

(1) *The Trademark Register.* Lists all trademarks registered in United States through 1976. Available from U.S. Trademark Association, 6 E. 45 St., New York, N.Y. 10019.

(2) *Registration and Use In the Trademark Laws of Different Countries.* Available from U.S. Trademark Association.

(3) *The Rocky Road Towards A European Common Market Trademark.* Available from U.S. Trademark Association.

(4) *Trademarks Throughout the World.* Available from Trade Activities, Inc., 347 Madison Avenue, New York, N.Y. 10017. Loose leaf and kept up to date with supplements.

COPYRIGHT

(1) Register of Copyrights, Library of Congress, Washington, D.C. 20540. They will furnish Application Form KK for packaging. A mandatory step for copyright registration.

Since the overall effect of copyright on packaging is rather negligible, books dealing with literary law are useful.

(2) Campbell, C., *The Business of Authorship.* T.V. Boardman & Co., Ltd. (U.K.) 1965.

(3) Paine, F. A., *Packaging and The Law.* Newnes-Butterworths (U.K.), 1973.

10

Canadian Regulations [1,2]

The first law of national scope in the Americas to protect the public against the adulteration of food, drink and drugs was the *Inland Revenue Act,* passed in 1875 by the Canadian government. This legislation was the result of pressure brought upon the government to check the adulteration of alcohol in a time when intemperance constituted an important social and health problem. The first annual report issued by the appointed analysts revealed that 51.5% of the food samples examined, mainly spices and milk, were adulterated.

A major amendment to this Act was passed in 1884 and the legislation was then known as the *Adulteration Act.* In 1910 the first standards for foods were promulgated after consultation with representatives from the food industry. The Adulteration Act was repealed in 1920 and superceded by the *Food and Drugs Act,* amended several times since then. The present Food and Drugs Act, passed in 1953, is administered jointly by Health and Welfare Canada and the Department of Consumer and Corporate Affairs. Major revisions in 1975 modified significantly the requirements for the labeling of foods. These amendments were coordinated with the requirements of the *Consumer Packaging and Labeling Act* promulgated in March, 1974, with manufacturers given until March, 1976, to comply with the new regulations. This Act is administered by the Department of Consumer and Corporate Affairs.

---

[1] Marvin A. Tung, Ph.D., P.Ag., Associate Professor, Department of Food Science, University of British Columbia, Vancouver, Canada, V6T 1W5.

[2] Maureen R. Garland, M.Sc., Senior Research Assistant, Department of Food Science, University of British Columbia, Vancouver, Canada, V6T 1W5.

The Consumer Packaging and Labeling Act and the Food and Drugs Act are the major pieces of legislation governing Canadian packaging and labeling, but other federal acts also contain relevant regulations. For subjects dealt with by the Consumer Packaging and Labeling Act, this Act is the overriding legislation. However, for matters different from those covered in the Consumer Packaging and Labeling Act, other legislation, for instance, the *Canada Agricultural Products Standards Act* and the *Textile Labeling Act,* because of their highly specialized nature, may be more demanding and thus contain the operative packaging and labeling regulations. The long term aim is one act, one authority. Eventually regulations pertaining to packaging and labeling in these scattered pieces of legislation will all be swept under the coverage of a new Consumer Packaging and Labeling Act.

Products falling under the jurisdiction of other acts become the responsibility of those departments charged with their administration. These departments, Health and Welfare Canada, Environment Canada and Agriculture Canada, are also responsible for interpreting and administering the labeling provisions of the Consumer Packaging and Labeling Act and the Food and Drugs Act which apply to their areas of interest. Of course, final authority for these interpretations rests with the courts.

Regulations under the various acts are constantly being reviewed and updated. Requests for changes in the regulations may arise from any sector: consumer groups, industry, and researchers. To keep up with new findings and to cope with the demands of modern industry, the legislative system must be flexible enough to allow for the introduction of new regulations without undue delay. Such a provision is contained in these acts whereby authority has been entrusted to the Governor-in-Council (that is, the Cabinet composed of the Prime Minister and Ministers of the Crown) for making new regulations; these have the same force and effect as the Act itself. Often proposals for modifications are first communicated to industry by means of an *Information Letter* which briefly outlines the proposed regulation and invites comments and discussion. The regulation is finally published in the *Canada Gazette* when it becomes law. Although there are no provisions for public hearings, the responsible Minister answers to Parliament and, thus, through the democratic process, to the electorate. Furthermore, the validity of any regulation is subject to challenge in the courts.

CONSUMER PACKAGING AND LABELING ACT

All prepackaged consumer goods, whether food or nonfood products, must conform to the requirements of the Consumer Packaging and

Labeling Act. The Act establishes authority for inspection, seizure, and detention. Products must meet packaging requirements and bear a label containing the following information in the specified languages and locations:

(1) *Net quantity of product.*—This must be clearly and prominently declared in Canadian (avoirdupois, U.K. or Imperial for volume) and metric units of measurement and in Canada's two official languages, English and French, on the principal display panel (PDP).
(2) *Identity of the product.*—The product identity, in terms of its common or generic name or in terms of its function, must be shown in English and French on the PDP.
(3) *Identity and principal place of business of the person by or for whom the prepackaged product was manufactured or produced for resale.*—This information must be sufficient for postal purposes and may be stated in either English or French. The declaration may be located on any part of the label except that part, if any, applied to the bottom of the container. Where the product was manufactured in a country other than Canada and the label shows the name and address of a person in Canada for whom it was manufactured, the name and address of that person must be preceded by "imported by" ("Importé par") or "imported for" ("importé pour"), unless the geographic origin of the prepackaged product is stated on the label.
(4) *Such other information required by the Act or Regulations to be shown on the PDP.*

There is no restriction on the style of type faces used except that the information required must be easily legible and in letters of not less than one-sixteenth in. in height, except when the PDP is less than 1.55 sq in. where letters not less than one thirty-second in. may be used.

For the purposes of the Act and Regulations "principal display panel" (PDP) means:

(1) that part of the label applied to all or part of the principal display surface:
  (a) in the case of a container that has a side or surface that is displayed or visible under normal or customary conditions for sale or use, the total area of such side excluding the top, if any;
  (b) in the case of a container that has a lid that is the part of the container displayed or visible under normal or customary conditions of sale or use, the total area of the top surface of the lid;

(c) in the case of a container that does not have a particular side or surface that is displayed or visible under normal or customary conditions of sale, any 40% of the total surface area of the container, excluding the top and bottom, if any, if such 40% can be displayed or visible under normal or customary conditions of sale or use;

(d) in the case of a container that is a bag with sides of equal dimensions, the total area of one of the sides;

(e) in the case of a container that is a bag with sides of more than one size, the total area of one of the largest sides;

(f) in the case of a container that is a wrapper or confining band that is so narrow in relation to the size of the product contained that it cannot reasonably be said to have any side or surface that is displayed or visible under normal or customary conditions of sale or use, the total area of one side of a ticket or tag attached to such container;

(2) in the case of a container that is mounted on a display card, that part of the label applied to all or part of the principal display surface of the container or to all or part of the side of the display card that is displayed or visible under normal or customary conditions of sale or use or to both such parts of the container and the display card;

(3) in the case of an ornamental container, that part of the label applied to all or part of the bottom of the container or to all or part of the principal display surface; and

(4) in the case of a label applied to a food that is not a prepackaged product, that part of the label applied to all or part of the side or surface of the food that is displayed or visible under normal or customary conditions of sale or use.

Where there are one or more surfaces on the label that are of at least the same size and prominence as the PDP, the information required may be shown in one official language if also shown in the other official language on one of the other surfaces. Examples of the mixed and separate language declarations are shown in Fig. 10.1.

The label must not contain any false or misleading representations relating to or that may be reasonably regarded as relating to that product. This includes such things as food products which connote a natural food flavor by means of a pictorial representation but do not in fact contain the natural food flavor. That the flavoring ingredient is imitation, artificial, or simulated must be declared in English and French, in type as large as that specified for net quantity declarations adjacent to the common name on the PDP if the pictorial representation is not on the PDP, or adjacent to the pictorial representation if it is on the PDP.

Used with permission of Quaker Oats Company of Canada, Ltd.

FIG. 10.1. EXAMPLES OF SEPARATE AND MIXED ENGLISH AND FRENCH LANGUAGE LABEL DECLARATIONS

Moreover, no representation may be made with respect to the number of servings contained unless that label contains a declaration of net quantity of each serving in the prescribed form and manner and in terms of numerical count, or acceptable unit of measurement.

Exemptions

Prepackaged products manufactured for export, for sale at a duty free store, or for commercial or industrial enterprises or institutions for their own use are exempted from all provision of the Act, as are products that are drugs or devices within the meaning of the Food and Drugs Act and prepackaged products consisting of replacement parts for vehicles, appliances, and other durable consumer goods, unless they are displayed to the consumer. Other products (feeds, fertilizers, pest control products, seeds, reusable container soft drinks, one-bite confections, fresh fruit and vegetables packaged in a band less than one half in. wide and prepackaged non-food products usually sold by numeric count and sold in such a manner that the contents are visible and identifiable or have a pictorical representation) are exempt from some major provisions of the Act.

There are also some exceptions to the dual language labeling requirement. Specialty products, greeting cards, books, talking toys, and games in which a knowledge of the language used is a basic factor essential to the use of the game need not be bilingually labeled. A local product or test market product is exempt if sold in a local government unit in which one of the official languages is the mother tongue of less than 10% of the local population and the information required is shown in the official language that is the mother tongue of at least 10% of the local population. Test market products are only exempt if six weeks notice of intention has been filed with the Minister of Consumer and Corporate Affairs.

Prepackaged individual servings of food that are served by a restaurant or prepared by a commissary and sold by automatic vending machines or mobile canteens and prepackaged catch weight products sold to a retailer by a manufacturer, processor, or producer are exempt from the net quantity declaration. When a product is declared by count and if there is only one, the net quantity may be declared by the use of the common name of the product in singular form. Only one of Canadian or metric units for the net quantity declaration is required as follows:

(1) where the net quantity is declared by count;
(2) where prepackaged products are packaged from bulk on retail premises;
(3) where prepackaged catch-weight products are sold by a retailer;
(4) where prepackaged products were being manufactured in Canada and labeled in terms of metric units only on the day the Act came into force;
(5) those prepackaged food items that are labeled in terms of whole numbers of kilograms or liters or in terms of the following units of grams or milliliters: 1-60, 125, 250, 500, 750;
(6) those prepackaged nonfood items that are labeled in terms of whole numbers of kilograms or liters or in terms of the following units of grams or milliliters: 1-60, 100, 150, 200, 250, 500, 750;
(7) those prepackaged products that are imported into Canada and are labeled in the country of origin in metric units only; and
(8) those prepackaged products that are packaged in standardized containers as specified in the Regulations (i.e., toothpaste, shampoo, toiletries, facial tissue) where metric declarations or numerical counts are required.

Net Quantity Declaration

The general rule for declaring net quantity is that the quantity is to be declared by volume when the product is a liquid or gas (except aerosols by weight) and by weight when the product is solid or viscous (i.e., honey, yoghurt). When it is established trade practice to show net quantity by numerical count or linear, area or cubic measurement, these units may be used. The manner of declaring net quantity for a number of specific products is detailed in the Regulations. For instance: canned shellfish by drained weight; canned fruits and vegetables by volume; frozen corn on the cob by numerical count; and waxed paper by the number of rolls, if more than one, length and width of each roll and total area.

Symbols and abbreviations may be used for the net quantity declaration providing these are as specified in the Weights and Measures Act.

Metric symbols must be shown in lower case letters while Canadian units may be in either upper or lower case. Table 10.1 presents acceptable units of measurement and their abbreviations. It is of interest to note that where a symbol is used to denote the word "liter," the use of the script "*l*" is mandatory.

If the full word (i.e., "gram") is used, it is necessary to show the equivalent word (i.e., "gramme") in French. It is not necessary to use such words as "Net," "Net Weight," "Net Contents," or "Net Quantity" as part of the declaration, but if used they must be shown in both English and French.

TABLE 10.1. ACCEPTABLE UNITS OF MEASUREMENT AND THEIR SYMBOLS

Units		**Symbols**	
English	French	English	French
gram	gramme	g	g
kilogram	kilogramme	kg	kg
litre	litre	ℓ	ℓ
millilitre	millilitre	ml	ml
centimetre	centimètre	cm	cm
metre	mètre	m	m
square centimetre	centimètre carré	cm^2	cm^2
square metre	mètre carré	m^2	m^2
cubic centimetre	centimètre cube	cc or cm^3	cc or cm^3
cubic decimetre	decimètre cube	dm^3	dm^3
cubic metre	mètre cube	m^3	m^3
ounce	once	oz OZ	oz OZ
pound	livre	lb LB	lb LB
fluid ounce	once fluide	*fl oz FL OZ	*oz fl OZ FL
	once liquide		oz liq OZ LIQ
pint	chopine	pt PT	chop CHOP
quart	pinte	qt QT	ptc PTE
gallon	gallon	gal. GAL.	gal. GAL.
inch	pouce	in. IN.	po PO
foot	pied	ft FT	pi PI
yard	verge	yd YD	vg VG
	yard		yd YD
square inch	pouce carré	sq in. SQ IN.	po^2 PO^2
square foot	pied carré	sq ft SQ FT	pi^2 PI^2
square yard	verge carré		vg^2 VG^2
	yard carré	sq yd SQ YD	yd^2 YD^2
cubic inch	pouce cube	cu in. CU IN.	po^3 PO^3
cubic foot	pied cube	cu ft CU FT	pi^3 PI^3
cubic yard	verge cube		vg^3 VG^3
	yard cube	cu yd CU YD	yd^3 YD^3

**An acceptable combined bilingual declaration may be shown as "oz fl" or "OZ FL."*

The numerical quantity in the declaration must be shown in bold face type in letters with heights not less than those specified in Table 10.2, depending on the area of the PDP.

One single type space should be used to separate the number from the unit of measurement. The use of an oblique stroke to separate Canadian from metric units of measurement is unacceptable; however, the units must be grouped together except that any symbol required to be shown pursuant to the Hazardous Products Act may be shown between and immediately adjacent to the units.

The Regulations further specify the units to be used. For instance, where the net volume of the product is less than 1000 ml, the metric units must be milliliters except that 500 ml may be shown as being one-half liter. Allowable tolerances for all net quantity declarations are fully specified.

There is also a provision in the Regulations directly aimed at eliminating any consumer confusion resulting from the difference between Canadian and U.S. volume measure. The Canadian measure is Imperial so that while a U.S. gal. is 128 oz, a Canadian gal. is 160 oz. Receptacles designed for household, camping, or recreation use (including water tanks), where labeled with a description of size or capacity in terms of any other pt, qt, or gal., must also be marked, adjacent and in equal type size, in terms of Canadian pt, qt, or gal.

TABLE 10.2. TYPE HEIGHT OF THE NUMERICAL QUANTITY IN THE NET QUANTITY DECLARATION[1]

Area of PDP		Height of Type Size	
sq in.	sq cm	in.	mm
< 1.55	< 10	1/32	0.8
< 5	< 32	1/16	1.6
5-40	32-258	1/8	3.2
40-100	258-645	1/4	6.4
100-400	645-2580	3/8	9.5
> 400	> 2580	1/2	12.7

[1] Type height means the height of an upper case letter where upper case is used and the height of the lower case letter "o" when lower case or a mixture of upper and lower case is used.

Standardization

"Where the Governor in Council is of the opinion that there is an undue proliferation of sizes and shapes of containers in which any prepackaged product is sold and that the effect of such undue proliferation of sizes and shapes is to confuse or mislead or be likely to confuse or mislead consumers as to the weight, measure or numerical count of a prepackaged product, the Governor in Council, on the recommendation of the Minister, may make regulations establishing packaging requirements whereby the size and shapes of container in which that prepackaged product or class of prepackaged product may be sold are limited."

Needless to say, this clause in the Consumer Packaging and Labeling Act is the most controversial. Industry has called the power of standardization of container sizes dictatorial suggesting it would limit the consumer's freedom of choice in the marketplace and stifle innovation. Consumer groups view the proliferation of sizes and prices as a marketing gimmick and welcome standardization. Standardization of container sizes is definitely underway with toiletries, the first products affected. Tube toothpaste, for instance, may only be sold in whole number sizes up to 25 ml, 50, 75, 100, 125, 150 ml and larger sizes that are multiples of 50 ml.

Metrication

To a large extent standardization of container sizes will come automatically with the metric conversion currently underway in Canada. The legislation regarding metrication is presently being enacted by the Canadian parliament. It is expected that major conversion will be accomplished by 1980. Working with industry, the Metric Commission has proposed standard metric package sizes for different types of food and deadlines for conversion. Proposed package sizes allow easy comparison between brands and provide the consumer with a reasonable choice of the amount of product purchased. Proposed sizes for instant coffee, for example, are 50, 200, 300 g with additional increments of 100 g; for sugar, 500 g, 1, 2 and 4 kg with 10, 20 and 40 kg institutional sizes. Already many products have converted. The dairy industry is ahead in this regard with milk now sold in liter containers and yoghurt, cottage cheese, etc., in "round number" ml or g containers.

Problem areas in the conversion are largely centered around rigid packs, particularly the tinplate can. Although most conversion will be "hard," that is with volumes or weights in "round number" metric sizes (i.e., 1 lb packs (454 g) replaced by 500 g packs), some conversions will be "soft," as in the baking industry where changing bread pans would be costly (the 16 oz loaf will be replaced by the 450 g loaf) and in the beer industry because of the uniform bottles presently used.

As metrication occurs it is expected, of course, that regulations requiring net quantity declarations in Canadian units of measure will be revoked. Indeed, recent advice from Ottawa makes it possible for all products to go completely metric without stating the Canadian measure, although both systems may be used.

FOOD AND DRUGS ACT

The intent of the Food and Drugs Act is to protect the public against health hazards and fraud in the sale and use of foods, drugs, cosmetics and medical devices. Health and Welfare Canada has responsibility for any health hazards while the Department of Consumer and Corporate Affairs is responsible for enforcing all aspects pertaining to economic fraud and labeling.

Deception is prohibited as the Act states no person shall label, package, treat, process, sell or advertise any food, drug or device in a manner that is false, misleading or deceptive or is likely to create an erroneous impression regarding its character, value, composition, merit, or safety. The Act specifically forbids sale of any food, drug, cosmetic, or device that is represented, either on the label or through advertising as a preventative, treatment or cure for a number of specific diseases ranging from alcoholism through leukemia to venereal disease. Some drugs and a device are exempted. Advertising of contraceptives as such is prohibited.

The Act prohibits sale of an article of food that:

(1) has in or upon it any poisonous or harmful substance;
(2) is unfit for human consumption;
(3) consists in whole or part of any filthy, putrid, disgusting, rotten, decomposed or diseased animal or vegetable substance;
(4) is adulterated; or
(5) was manufactured, prepared, preserved, packaged, or stored under unsanitary conditions.

Many of these general prohibitions apply also for drugs and cosmetics. For the latter, there is the additional requirement that no person shall sell any cosmetic that has in or upon it any substance that may cause injury to the health of the user when used according to directions on the label or for such purposes and by such methods as are customary. For devices, there is a similar prohibition related to causing injury.

The Act specifies that a food is considered adulterated if certain substances or classes of substances are present or have been added to the food. These include such things as mineral oil and paraffin wax (with some low level exceptions), coumarin, safrole, etc. Permissive limits are set for heavy metals and for many other "adulterants"; many of the set tolerances differ for various foods.

Use of a few food packaging materials is also regulated. Here the intent is that no person shall sell any food in a package that may yield to its contents any substance that may be injurious to the health of a consumer of the food. Beyond this stipulation, however, the Act does not contain the detailed regulations governing packaging components and migration and extraction tests found in U.S. regulations.

Another major thrust of the Act concerns adherence to standards of identity. For articles of food, drugs and cosmetics, standards of composition, strength, potency, purity, quality, or other properties are prescribed. Labeling, packaging, selling, or advertising any article in such a manner that it is likely to be mistaken for a standardized product unless it complies with the standard is forbidden. For drugs, this includes those for which a standard has not been prescribed but for which a standard is contained in specified major pharmaceutical publications. Some differences in color, flavor, shape, and size are permitted for drugs if the difference does not interfere with assay methods. Certain drugs, such as thalidomide, are specifically banned from sale.

The Food and Drugs Act provides for inspection, forfeiture, analysis, evidence, and, of course, penalties to back up its intent. Moreover, for drugs, it specifies restricted and controlled drugs and the limitations for possession and trafficking in these drugs.

Food Regulations

With the exception of one-bite confections and fresh fruits and vegetables packaged in a wrapper less than one-half in. wide, all prepackaged food and a few other foods such as barbecued poultry or meat prepared by the retailer must bear a label when it is sold. The label must be applied to the package in the case of a prepackaged food or to the food itself.

Besides the declarations of common name, net weight, and name and address of manufacturer as required by the Consumer Packaging and Labeling Act, the label must declare, in both English and French:

(1) where the durable life of a prepackaged product is 90 days or less,
   (a) the durable life date, and
   (b) instructions for the proper storage of the prepackaged product if it requires storage conditions that differ from normal room storage;

(2) a list of all ingredients, including components, if any, where the prepackaged product consists of more than one ingredient; and

(3) any other information as required by the Regulations.

The latter declaration includes such statements as "previously frozen" for meat, poultry, or fish and their products which have been frozen and then thawed prior to packaging, "simulated" and "contains no meat/poultry" as the case may be for simulated products, declarations of extenders, vitamin additions, etc., detailed requirements for which are given in the Regulations.

As in the Consumer Packaging and Labeling Act there are a few exemptions from the dual language requirement. Label locations for some of the required information are specified. It is not to be shown on that part of the label, if any, applied to the bottom of a food or container unless the container is ornamental or the information is also shown on parts of the label that are not applied to the bottom or if a clear indication of the location of the required information appears elsewhere on the label.

Used with permission of Campbell Soup Co., Campden, N.J., Campbell Soup Co., Ltd., Toronto, Ont.

FIG. 10.2. LABELS OF CANADIAN (TOP) AND AMERICAN (BOTTOM) VERSIONS OF THE SAME PRODUCT

It is of interest to compare the Canadian and American labels of a product (Fig. 10.2). The Canadian label, as a result of the dual language labeling requirement, is much "busier"; the required declarations in English and French leave no room for recipes.

Durable Life Date.— Products with a durable life of less than 90 days must show the durable life date in the following manner. The words "best before" and "meilleur avant" are to be shown grouped together with the durable life date unless a clear explanation of the significance of the date appears elsewhere on the package. The date itself shall show the year, represented by at least the last two numbers of the year, where the year is necessary for the sake of clarity, then the month followed by the day of the month. The month may be abbreviated if necessary. The following are acceptable abbreviations for both the English and French languages:

JA for January
FE for February
MR for March
AL for April
MA for May
JN for June
JL for July
AU for August
SE for September
OC for October
NO for November
DE for December.

Courtesy of Tung and Garland

FIG. 10.3. EXAMPLES OF CANADIAN DURABLE DATE DECLARATIONS

Products such as individual portion packs served by restaurants or vending machines are exempt from durable dating.

It is of interest to note that instead of "best before" the phrase "use before" was originally proposed for the Regulations — but, after extensive study and consultation with consumer and trade associations, the Consumer and Corporate Affairs department amended the proposed regulation to "best before." This phrase more properly implies to consumers that, if properly handled, the product should be of high quality until the specified date, but that the product may still be consumed beyond the date providing it is in an acceptable state.

Ingredient Listing.— Ingredients must be shown in descending order or percentage of their proportions of the prepackaged product before they were combined to form the product, except that:

(1) spices, seasonings, and herbs, except salt;
(2) natural and artificial flavors;
(3) flavor enhancers;
(4) food additives, except ingredients of food additives or mixtures of substances for use as a food additive;
(5) vitamins;
(6) salts or derivatives of vitamins;
(7) mineral nutrients; and
(8) salts of mineral nutrients

may be shown at the end of the list of ingredients in any order. Components must be shown immediately after the ingredient of which they are components in such a manner as to indicate they are components of that ingredient and in descending order of their proportion in that ingredient. All this information must be shown grouped together in the same official language.

A number of foods such as those packaged from bulk and poultry or meats barbecued on the retail premises, individual serving packs sold by restaurants, canteens, and vending machines as well as many products subject to compositional standards (i.e., alcoholic beverages, vinegars) are exempted from ingredient declarations. A great number of standard ingredients (i.e., margarine, rice, baking powder, spice mixtures) are exempted from the ingredient components declaration. However, such ingredient components as flavor enhancers, salt, and food additives and all those used in excess of amounts for normal effects must be shown. Where a vitamin or mineral nutrient is a component of any ingredient and if a reasonable daily intake (RDI) of that food would result in a daily intake of more than various specified levels or if it is present in amounts equal to or in excess of the amount set out in the compositional standard, then the vitamin or mineral must be shown on the label. Otherwise, it need not be shown.

Ingredients are to be listed under their common name although certain ingredients may be listed under a class name (i.e., various vegetable oils and fats may be listed as "vegetable oil" or "vegetable fat"). Meat, poultry, fish or their by-products must be identified under their individual names. In the case of plant protein products and protein isolates, the name of the source of the product must be given.

If an ingredient is optional or can be substituted for another one in the preparation of a product, the label may indicate all the ingredients that are likely to be used in this product during one year. However, there must be a clear indication that these specific ingredients may not all be present in a given package of food.

A recent (August 5, 1977) Information Letter to industry from the Health Protection Branch is of particular interest as it proposes to amend the Regulations so that the total content of sugar and other sweeteners must be declared as a percentage of the total weight of the cereal on the PDP of all breakfast cereals. This declaration would be based on the total amount of hexoses and disaccharides in the product as sold and would take the form "contains (x)% sugar (and other sweeteners)." The comments of industry may, of course, influence this proposed legislation.

Dietary Claims.— The Act provides specific Regulations for statements or claims in advertisements or on the label which imply a special dietary use. To avoid the proliferation of products recommended for all sorts of ailments, the new Regulations clearly specify the types of products that may be marked as "dietetic." There are two permitted product groups:

(1) Products for which there are no present compositional or labeling requirements but which can be recommended for the following diets:
   gluten-free,
   protein-restricted,
   elemental or chemically defined,
   tube feeding; and

(2) products that must meet the compositional and labeling requirements set for one of the following foods:
   carbohydrate-reduced,
   sugar-free,
   calorie-reduced,
   low calorie,
   low sodium.

Besides general labeling requirements the labels of these latter foods must declare:

(1) the type of diet the product is recommended for;
(2) the appropriate expression, carbohydrate-reduced, sugar-free, etc., in close proximity to the common name of the food and in the same size type;
(3) the content of carbohydrate, protein, and fat and the content of mannitol or sorbitol when used; and
(4) the calories expressed per unit of ready-to-serve food and per 100 g or 100 ml.

The label of a low sodium food must also declare the sodium and potassium content per unit on a ready-to-serve or mg per 100 g or 100 ml basis.

The Regulations state that a carbohydrate-reduced food may contain not more than 50% of the available carbohydrate found in the food, may not contain more calories than the food it replaces and may be made only providing the original product was a significant source of carbohydrate and derived at least 25% of its calories from carbohydrate. A sugar-free food is defined as a carbohydrate-reduced food containing not more than 0.25% available carbohydrate and may provide not more than one calorie per 100 g or 100 ml except for chewing gum. Sugar-free foods may be described as non-cariogenic or any synonymous term. A calorie-reduced food must contain not more than 50% of the calories found in the food it replaces as must a low calorie food except that it must also supply not more than 15 calories per average serving and 30 calories in a RDI. A low sodium food cannot contain more than 50% of the sodium present in the food for which it is a substitute, must supply not more than 40 mg Na/100 g (80 mg Na/100 g for fish, meat and poultry) and may not contain added Na salt. Further guidelines are given for other dietary labeling statements and dietetic versions of standardized foods.

Protein may not be mentioned in the common name of a food unless that food is described on the label as "an excellent source of protein," "an excellent dietary source of protein," "a good source of protein," "a good dietary source of protein," or that it contains "high quality protein" for which statements the protein must meet defined Food and Drugs Regulations criteria. Limited claims such as "proteins help provide food energy" may be made on the label if the food is described as an "excellent" or "good" source of protein. Protein may be mentioned in the list of ingredients when used in the formulation (i.e., hydrolyzed vegetable protein) and the amount of protein, fat, and carbohydrate may be stated if the three statements appear together and receive equal prominence.

The only claim permitted in the advertisement and labeling of fats and oils is a statement of the percentage by weight of polyunsaturated and saturated fatty acids in the total fats providing the saturated fatty acids do not exceed 20% and the polyunsaturates comprise at least 40% of the fat if it is an oil, or at least 25% in shortening or similar products. These statements must be grouped together, be given equal prominence and clearly refer to total fat. Any other reference to fatty acids or cholesterol is prohibited.

Enrichment.— Additon of certain nutrients to food has been controlled in Canada for more than a quarter of a century. As a result of nutrition surveys in Newfoundland in the forties which revealed widespread vitamin deficiencies, the Food and Drugs Regulations were amended to permit voluntary enrichment of white flour and bread with thiamin, riboflavin, niacin, and iron. With time, the criteria for food enrichment were further revised and updated. Today, the enrichment of food is permitted or made mandatory when there is *nutritional justification,* generally:

(1) when there is insufficient intake of a nutrient, as in the case of iodine where the addition of 0.01% potassium iodide to table salt is mandatory and the presence of iodide must be indicated on the PDP;

(2) when a product replaces a meal or food normally considered an important source of essential nutrients, for instance, margarine, infant formula, meat and poultry substitutes and extenders and instant breakfasts where, for the latter, the Regulations state, "no person shall sell a product represented as a ready breakfast or instant breakfast or by any similar designation unless each portion or serving of the product contains: (a) not less than 4.0 mg iron; (b) Vitamin A, thiamin, riboflavin, niacin or niacinamide, and Vitamin C; (c) a good dietary source of protein; and (d) where consumed as directed, not less than 300 calories"; and

(3) when a vitamin or mineral is removed from a staple food in the course of good manufacturing practice, as in flour.

To ensure appropriate enrichment, the Regulations specify foods which may be enriched and specify the minimum and maximum amounts, generally 40-70% and 100-200% respectively of the RDI, of certain nutrients which may be added. New nutritional requirements for breakfast cereals are currently being proposed. When a food is enriched, the amount of vitamin, mineral or amino acid added to it must be stated on the label.

Claims based on the vitamin and mineral content of food are regulated to prevent exploitation. The use of the terms "excellent" or "good"

Used with permission of Kellogg Co. and Kellogg Salada Canada Ltd.

FIG. 10.4. THE EFFECT OF DIFFERENCES IN ENRICHMENT POLICY ON THE FORTIFICATION OF AMERICAN (LEFT) AND CANADIAN VERSIONS OF A PRODUCT.

sources of a vitamin or mineral is restricted to foods with a naturally occurring vitamin or mineral content. In such cases the claim or more general statements such as the vitamin or mineral is a "factor in the maintenance of good health" are permitted if the RDI of the food provides specified amounts of the nutrient. There are further restrictions regarding claims for foods intended for children.

Differences between Canadian and U.S. approaches to regulating enrichment are reflected in nutritional statements on the labels of Canadian and American versions of the same product (Fig. 10.4).

Standards.— For approximately 300 food items on the market, the Food and Drugs Regulations define a standard of identity or composition which ensures that all products sold under a given name meet a certain quality. The Regulations specify that where a standard is prescribed for a food, the food shall only contain the ingredients included in the standard for the food and that each ingredient must be incorporated within the prescribed quantity limits and that any food additive must be one specified and approved for use in that food.

Standards are prescribed for alcoholic beverages, including regulations regarding alcohol volume, aging, and country of origin declarations; cacao products; coffee; food colorings; spices and seasonings; dairy products; fats and oils; flavoring preparations; fruits and vegetables and their products; grain and bakery products; meat preparation and products and meat extenders; sweetening agents; salt; vinegars; marine and fresh water animal products; poultry preparation and products; and egg products including liquid egg. The ingredients and amounts which constitute the product are carefully defined. For example, mayonnaise,

"(1) shall be a combination of
  (a) vegetable oil,
  (b) whole egg or egg yolk, in liquid, frozen or dried form, and
  (c) vinegar or lemon juice;
(2) may contain
  (a) water,
  (b) salt,
  (c) a sweetening agent,
  (d) spice or other seasoning except tumeric or saffron,
  (e) citric, tartaric or lactic acid,
  (f) a sequestering agent; and
(3) shall contain not less than 65 percent vegetable oil."

Salad dressing, on the other hand,

"(1) shall be a combination of
(a) vegetable oil,
(b) whole egg or egg yolk, in liquid, frozen or dried form,
(c) vinegar or lemon juice, and
(d) cereal;
(2) may contain
(a) water,
(b) salt,
(c) a sweetening agent,
(d) spice, or other seasoning,
(e) an emulsifying agent,
(f) citric, tartaric or lactic acid, and
(g) a sequestering agent; and
(3) shall contain not less than 35 per cent vegetable oil."

For dairy products the standards include Regulations making vitamin additions to milk mandatory (except for products to be used in manufacturing) and Regulations regarding fat content declarations.

Where meat, poultry or fish and their byproducts that have been frozen are thawed prior to sale, the words "previously frozen" must be prominently shown. Where a part of a food has been frozen and thawed "made from fresh and frozen portions" must be in evidence on the label.

A meat or poultry extender must be labeled with the common name of the food in the extender that is a source of protein, plus the word "meat," "poultry," or the common name of the product to be extended, plus the word "extender." The words "contains no meat" or "contains no poultry" must be shown on the label of a simulated product in close proximity to the common name which must be modified by the word "simulated." All these qualifications must be in letters of at least the same size and prominence as the common name. Enrichment is mandatory where required for the product to meet specified nutritional levels and the words "(when rehydrated as directed) contains approximately the same amount of vitamin or mineral nutrients as an equal weight of meat (poultry)" must be shown on the label.

Additives.— According to the Regulations, a food additive is "any substance, including a source of radiation, the use of which results, or may be reasonably expected to result in it or its byproducts becoming a part of or affecting the characteristics of a food." Health and Welfare Canada states that, by this definition, a food additive does not include:

(1) any nutritive material that is used, recognized or commonly sold as an article or ingredient of food;

(2) vitamins, mineral nutrients, and amino acids;
(3) spices, seasonings, flavoring preparations, essential oils, oleoresins, and natural extractives;
(4) agricultural chemical residues;
(5) food packaging materials and components thereof; and
(6) drugs recommended for the administration to animals that may be consumed as food.

Other regulations govern the use of these products.

Where a standard for a food is not prescribed, the food may only contain certain food additives specified for use as additives to that food for set purposes and may be incorporated only in a quantity within specified limits. Where the limit is stated as "good manufacturing practice" the amount must not exceed that required to accomplish the purpose for which the additive is permitted. For unstandardized foods, food additives used in the product must be declared on the label. Additives that may be used in infant foods are severely limited.

The food additives must meet purity criteria set out in the Regulations or, where not prescribed, those specifications set out in publication 1406 "Food Chemical Codex — Second Edition," published by the National Academy of Science, National Research Council of U.S.A.

The Regulations provide tables of accepted food additives and their maximum level of use, classing them as coloring agents; anticaking agents; bleaching, maturing, and dough conditioning agents; emulsifying, gelling, stabilizing, and thickening agents; food enzymes; firming agents; glazing and polishing agents; non-nutritive sweetening agents; pH adjustment agents; acid-reacting materials and water correcting agents; Class I-IV preservatives; sequestering agents; starch modifying agents; yeast foods; carrier or extraction solvents; and miscellaneous food additives. No more than one of the following:

(1) benzoic acid and its salts;
(2) methyl-p-hydroxybenzoate and propyl-p-hydroxybenzoate and their salts; and
(3) sulfurous acid and its salt

may be used in or upon a food.

A food additive may be added to the list on submission of details (complete description, amount to be used, purposes, physical or technical effects, residues, safety tests, data establishing its intended effect, etc.) to the Food Directorate of the Health Protection Branch, Health and Welfare Canada. The criteria of evaluation include that the food additive must be safe for continued use, its use must not lead to deception and its use must result in an advantage to the consumer by improving or maintaining the nutritive value, quantity, quality or acceptability of the food.

The Minister will, within 90 days after the filing of the submission, notify the applicant as to whether or not he intends to recommend to the Governor-in-Council that the food additive be listed.

Drug Regulations

Drugs may not be sold unless they are labeled as required by the Food and Drugs Regulations. Besides the manufacturer's name and address, the inner and outer labels must bear, in both official languages:

(1) the name of the drug — drugs for which a standard is prescribed must show the proper name of that drug and state that the drug is a Canada Standard Drug (the abbreviation C.S.D. may be used) or the standard that the drug professes to meet. Where there is a proprietary or brand name for a drug, the proper name shall immediately precede or follow the proprietary or brand name in type not less than one-half size of that for the brand name. Where no standard is prescribed in the Regulations or specified publications, only the proper name, or if it has no proper name, the common name is specified;
(2) the lot number of the drug, preceded by Lot Number, Lot No., Lot, or L.;
(3) adequate directions for use of the drug;
(4) a quantitative list of the medicinal ingredients of the drug by their proper names, or, if they have no proper names, their common names;
(5) for drugs sold in dosage form, the drug identification number assigned to that drug, preceded by GP (for proprietary medicines) or DIN for others;
(6) for drugs which do not maintain potency, purity or physical characteristics for at least three years from the date of manufacture, an expiration date, which often must be within controlled limits;
(7) where the drug is intended for parenteral use, a quantitative list of any preservatives present therein by their proper names, or, if they have no proper names, their common names;
(8) where dose limits are specified in the Regulations,
  (a) the quantitative content of the drug,
  (b) the recommended single and daily adult dose designated as such, except for
    (i) preparations solely for external use, or

(ii) preparations solely for children's use, and

(c) adequate directions for use when the drug is recommended for children which shall be either

(i) the statement "CHILDREN: As directed by the physician," or

(ii) a suitably reduced, maximum single and daily dose which shall not exceed the following:

Age in years	Proportion of adult dose
10-14	one-half
5-9	one-fourth
2-4	one-sixth
under 2 years	as directed by physician

and;

(9) the net amount of the drug in the container in terms of weight, measure or number.

The last declaration, for which the Regulations specify tolerance limits, may be only on the outer label. All other declarations must be on both inner and outer labels unless the package has only one label where all required information must be on that label.

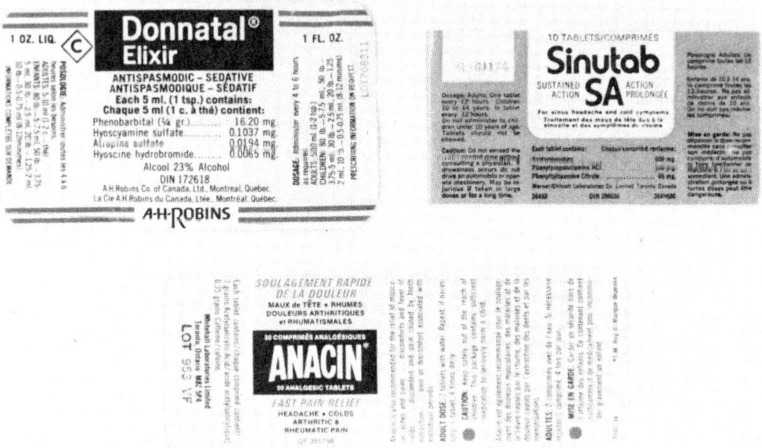

Used with permission of A.H. Robins Co. of Canada, Ltd. Warner-Chilcott Laboratories Co. Ltd. and Whitehall Laboratories Ltd.

FIG. 10.5. EXAMPLES OF CANADIAN DRUG LABELS
Note that the top left label is from a controlled drug.

Drugs sold to a manufacturer, packed from bulk on the retail premises, sold on prescription, and the inner label of single-dose container drugs are exempt from these labeling requirements. However, when a package of a drug bears a statement, mark or device regarding the ingredients therein and the drug name, name and address of the retailer, net contents and adequate directions for use, then it must be fully labeled as required in the Regulations.

Where a standard or grade is prescribed for a drug, then the name or designation for that drug may not be used unless the drug conforms to that standard or grade. Drugs with labels carrying concentration or dosage recommendations above the set limits must bear a caution that the product is to be used only on the advice of a physician. For specified drugs other cautionary statements (i.e., "keep out of reach of children," "shake well before use," "keep refrigerated") are required on the labels. Package content limits are often set. The Regulations specify form and color of the tablet for some drugs and that the immediate container must be readily distinguished by touch. Some drugs, for example, disinfectants, have further labeling requirements. The Canadian Licence number of the manufacturer may be required to be stated as may the name of the solvent or vehicle used in products in liquid form. Manufacturers of "time-release" drugs must be able to demonstrate that the drug is released and available as represented. These "time-release" drugs are exempted from tablet disintegration time specifications that must otherwise be met by any drug sold in tablet form. The use and levels of vitamins as well as claims made for vitamins in drugs are strictly regulated.

Antibiotics must declare, on both the inner and outer labels, the potency expressed in terms of International Units or, where this has not been established, other appropriate units and the expiration date.

Packaging materials must be examined and tested to establish that they are suitable for the intended use. Carton dimensions are specified for drugs sold in collapsible tubes with larger package dimensions permitted when an applicator is included in the carton provided the main label panel indicates it is a combination package.

Certain drugs cannot be advertised as being for sale to the public unless they are in a form not suitable for use in humans and are advertised and marked "(For) Agricultural Use Only" or "(For) Veterinary Use Only."

Labels of controlled drugs must bear a conspicuous prescribed symbol denoting this fact. When dispensed by a pharmacist, general labeling requirements do not apply but the label of the controlled drug package must state:

(1) the name and address of the pharmacy or pharmacist;

(2) the date and number of the prescription;
(3) the name of the person for whom the controlled drug is dispensed;
(4) the name of the practitioner;
(5) directions for use;
(6) any other information that the prescription requires to be shown on the label.

Every restricted drug sold to an institution must be securely packaged by the licensed dealer so that the package cannot be opened without breaking the seal. General labeling regulations do not apply. Inner and outer labels must show the proper or, if there is no proper name, the common name of the drug, net contents, unit strength where the drug is in unit form, lot number, name and address of the manufacturer, and the words "Restricted Drug."

Cosmetics and Devices Regulations

These reinforce basic Consumer Packaging and Labeling Act requirements with provisions that the labels must also bear adequate directions for safe use where a hazard exists. Required warnings, cautions, or special directions are specified. The Regulations limit uses or concentrations of certain cosmetic ingredients and specify the cautionary statements which must appear when these ingredients are used. Labels of deodorants intended for use in the genital area, for example, must have statements of caution as well as directions for use.

Labeling or advertising claims made respecting an action or effect of a cosmetic or any ingredient may not be made unless they are generally recognized as proper and are supported by adequate and proper tests. Claims implying the cosmetic has been prepared or compounded according to a prescription are not allowed.

For devices, the label must declare the name and address of the manufacturer, the name and lot number of the device, a list of contents of the package, and number of complete units contained within, as well as adequate directions for use.

HAZARDOUS PRODUCTS ACT

This Act and its Regulations prohibit the advertising, sale and importation of hazardous products except as authorized and providing they are packaged and labeled as prescribed. One of the main aspects of this legislation concerns standards which must be met to prevent hazards in a wide range of products, from safety glass through electrical appliances to fingerpaints and babies' pacifiers. Of interest in this discussion are certain packaging criteria and the required symbols, words, and statements denoting hazard which must be on the labels of restricted products.

The degree and nature of hazard symbols to be used are depicted in Figs. 10.6 to 10.8. Exact reproductions of these symbols must be clearly and prominently displayed, easily legible and in distinct contrast to other matter on the product container. They must also be readily discernible under customary conditions of purchase or use.

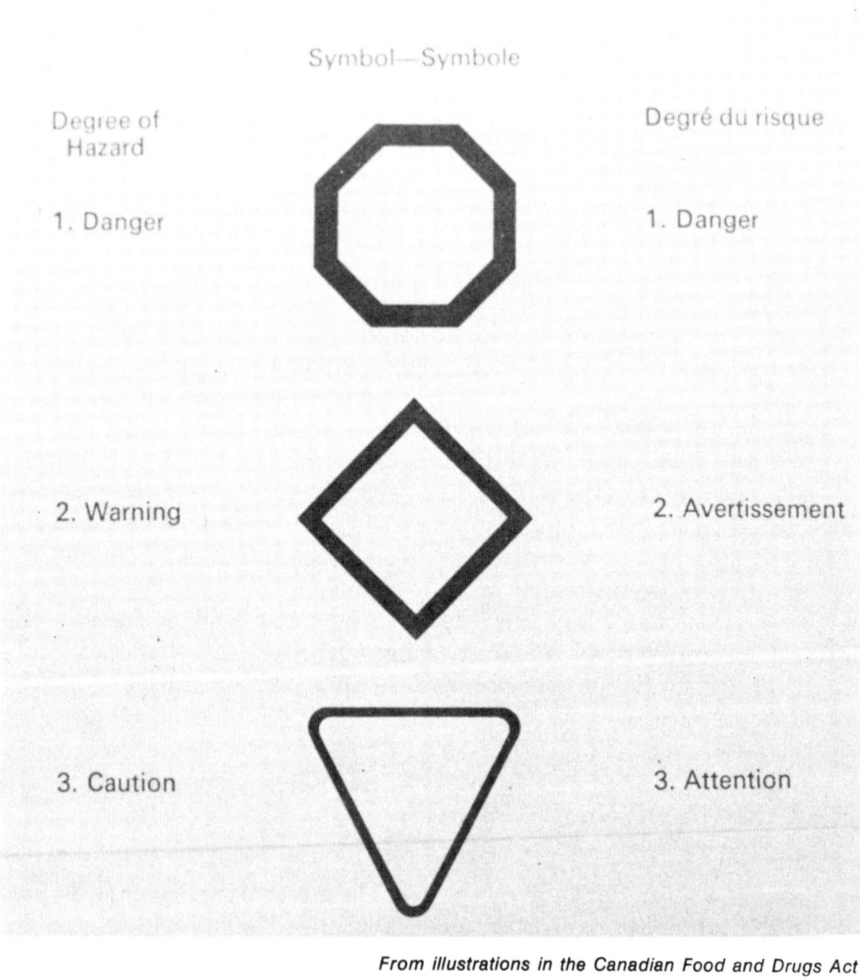

From illustrations in the Canadian Food and Drugs Act

FIG. 10.6. DEGREE OF HAZARD SYMBOLS

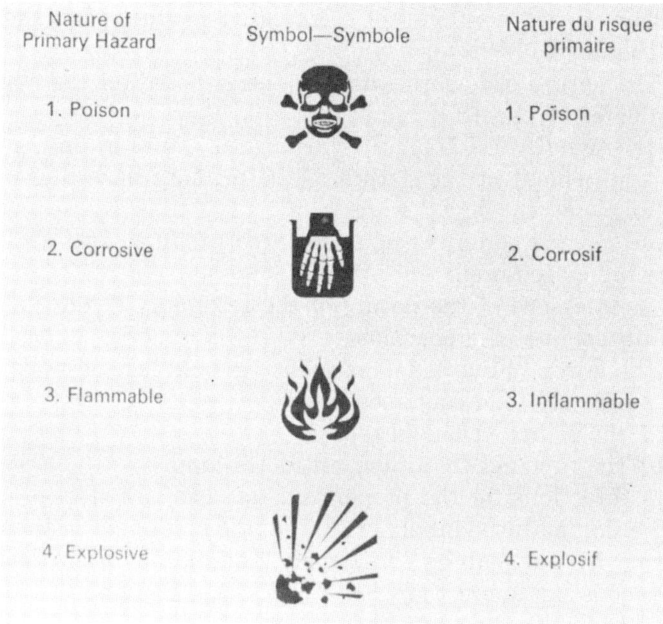

From illustrations in the Canadian Food and Drugs Act

FIG. 10.7. NATURE OF PRIMARY HAZARD SYMBOLS

From illustrations in the Canadian Food and Drugs Act

FIG. 10.8. REQUIRED RATIO OF SIZE OF NATURE OF HAZARD SYMBOL TO DEGREE OF HAZARD SYMBOL.

The PDP of each container of a regulated product shall show:

(1) the degree of hazard symbol;
(2) the nature of hazard symbol superimposed on the appropriate degree of hazard symbol;
(3) the signal word Danger, Warning, or Caution; and
(4) a statement of the nature of the primary hazard immediately following the signal word.

In addition each container of a regulated product must show such of the following as required:

(1) a statement of the nature of the secondary hazard;
(2) statements of precaution;
(3) the words "First Aid Treatment" in bold face capitals and, immediately following such words,
 (a) the source of the hazard
 (b) the first aid treatment antidotes, and
 (c) directions for the proper and safe use and storage.

Where the flammable symbol is required, the statement of the nature of the primary hazard shall be "Extremely Flammable" if the flashpoint of the product is not more than −6.7°C(20°F), "Flammable" if −6.7 to 26.7°C(20 to 80°F).

Statements must, of course, be in both English and French. Acceptable sizes of the symbols, signal words, nature of the primary hazard, and other statements are specified in the Regulations.

Hazard labeling requirements are stipulated for the following products: bleaches; cleansers and sanitizers containing chlorine; corrosive chemicals; petroleum distillates or products thereof; adhesives, solvents and dyes containing tolene or acetone; polishes, cleaning agents, liquid coating materials, paint and varnish removers; fire extinguisher fluids; antifreeze preparations; turpentines, methyl alcohol or products containing methyl alcohol; and aerosol containers of consumer products. For the latter, where the container does not contain a regulated product, the label must include:

(1) the Explosive symbol superimposed on the Caution symbol and the signal word "Caution";
(2) the nature of the primary hazard as "Container may explode if heated";
(3) the nature of the secondary hazard as "Contents under pressure"; and
(4) the statements of precaution "Do not place in hot water or near radiators, stoves or other sources of heat" and "Do not puncture or incinerate container or store at temperatures over 120°F."

Aerosol containers of regulated products naturally require further warnings, dependent on the product.

Moreover, for some products, use of child-resistant containers is mandatory; these must bear instructions in both the English and French languages on how to open and close the container. There is, however, the proviso that when such products are sold in containers of two or more different sizes, the containers of one size need not be child-resistant if they are bilingually labeled on the PDP with "this package is for households without children." Criteria for determining child-resistance are provided in the Regulations.

In the case of toys, equipment and other products for use by a child in learning or play that are packaged in flexible film bags, the flexible film bag must have an opening less than 14 in. in circumference, be made from film at least 0.75 mil thick and have printed on it the following warning: "Plastic bags can be dangerous. To avoid danger of suffocation, keep this bag away from babies and children" or different words conveying the same warning. Certain other products (i.e., chemistry sets, charcoal) must also bear warnings.

TEXTILE LABELING ACT

Consumer articles made in whole or part from textile fibers, yarns, or fabrics are required to be labeled according to the Textile Labeling Act and its Regulations. Articles such as clothing, wigs, floor coverings, sleeping bags, umbrellas, tents, etc., fall under the requirements of this Act and are thus exempt from the requirements of the Consumer Packaging and Labeling Act.

This Act refers to two types of labels: a "representation label" which contains any representation as to the textile fiber content of the article and a "disclosure label" which is a representation label fully complying with the labeling requirements of this Act and its Regulations and, where the information required is shown in two labels, means both. With some exceptions, representation labels are required to meet the criteria of disclosure labels which must declare, in English and French:

(1) the textile fiber content and percentages of each;
(2) the name and postal address of the dealer by or for whom the article was made or by whom the article was imported and labeled; and
(3) the name of the country of origin if there is representation that the article or any fiber or fabric therein is imported unless both the representation and declaration of country of origin is shown on another label.

A dealer who, pursuant to the Regulations, has an identification number from the Minister, may show this number in place of the name and address.

Textile Fiber Content

A number of articles such as those intended for one-time use only, shoes, luggage, home decorating items, sports equipment, lawn furniture, baby equipment such as playpens, belts, etc., and articles sold outside Canada or at a duty-free store are exempt from the textile fiber content declaration.

For other articles, each textile fiber that is present in the article or constituent part in amount of 5% or more by weight must be shown:

(1) by its generic name (as defined in the Regulations);
(2) in order of predominance by weight; and
(3) with the amount that is present preceeding or immediately following the generic name of the fiber.

Fibers present at less than 5% must be shown as "other fiber" with their amount. Where this fiber is present for a specific functional purpose or as ornamentation it may be shown by its generic name if the purpose for which it is present is also stated immediately before or after. The words "all" or "pure" may be used in place of words or figures indicating 100% when one textile fiber constitutes the entire fiber content of an article. A 5% tolerance in declared fiber amount is permitted.

In the case of reclaimed fibers, the word "reclaimed," "reprocessed," or "reused" must immediately precede the generic name of the fiber. A fiber trademark registered in Canada for that textile fabric and a descriptive term that is true in respect of that trademark may be shown immediately before the generic name. When textile articles are made up of two or more sections, other than findings, that differ in fiber content, each section must be named and described separately with its respective textile fiber contents. Further regulations specify fiber content declarations for articles that have linings, pile backings, trimmings, paddings, stuffings, or which contain findings. Constituent labeling requirements for fibers that are coextruded or physically combined (multiconstituent, matrix-fibril fibers) are also specified.

The information required by this legislation must be clearly and prominently shown in type of equal size and prominence, in one place, and set apart from any graphic material in the label which is not required. There are exceptions such as when labels have both sides readily accessible or for certain items where a separate label may be used. Acceptable label or declaration locations are also specified. False and misleading representations are, of course, prohibited. Use of words such as virgin, pure, hand-knitted or expressions, words, or depictions which denote fur, hair, or wool must be legitimately justified.

Voluntary Textile Labeling

Two textile labeling programs in Canada are optional on the part of manufacturers. Most have elected to comply.

The first of these is the Canada Standard Sizes (CSS) program for children's clothing. Garments carrying CSS labels are standardized so any one size is the same regardless of manufacturer. These sizes are based on body measurement and not the child's age.

The second program is care labeling. Labels may be attached to textile articles giving care instructions in colored symbols. A set of five symbols has been developed for washing, bleaching, drying, ironing, and dry-cleaning. They use a simple traffic light color code system: red means stop or don't; amber means proceed with caution; and green means go ahead (Fig. 10.9). Manufacturers are encouraged to use these care symbols on permanently attached labels so that the consumer has information which may influence his purchase choice and satisfactory results when cleaning the article.

OTHER FEDERAL FOOD LEGISLATION

Many other pieces of federal legislation govern the sale of food products in Canada. The main import of these Acts is protection of the public from health hazards through registration of establishments, inspection, and standards for food products. The Acts often specify packaging standards and requirements and, besides reinforcing the labeling requirements of the Consumer Packaging and Labeling Act and the Food and Drugs Act, demand declarations of grade, preparation methods, quality, etc., which are of relevance in this discussion.

Fish Inspection Act

This Act and its Regulations control products in international and interprovincial trade and are administered and enforced by the Fisheries Branch of Environment Canada with the Consumer and Corporate Affairs Department responsible for monitoring labels of marine products at

From illustrations from the Consumer and Corporate Affairs Canada publication "Look at That Label!"

FIG. 10.9. CARE LABELING SYMBOLS FOR TEXTILES
The symbols are color coded with red for stop, amber for be careful and green for go ahead.

the retail level. Fills for various can sizes and quality standards for marine products are specified. No container or label may bear a quality designation unless a standard for quality has been specified in the Regulations and the fish in the container meets that standard.

Labeling requirements include that, for canned fish, the common name of the fish must indicate, in both official languages, whether the fish has been prepared:

(1) by mincing, flaking or other special process;
(2) from selected parts of the fish; or
(3) for dietetic use.

Canned salmon must have the species declared as part of the common name, i.e., Sockeye Salmon. For fish, other than canned fish, every container or label must be bilingually marked with the grade, size, class, count, and moisture content as may be specified by the Regulations. Frozen Atlantic smelt, for example, must have a declaration as to the size of the fish.

Moreover, many of the standards require label markings denoting quality or grade such as "light meat" tuna, "fancy shape" oysters, "dressed" pickled herring, "small" spring mackerel, and "choice" light salted cod. Canned salmon of the species known as Spring, King, or Chinook must be designated as "red," "pink," or "white," to indicate the color according to standards. Sockeye or Coho salmon that are excessively pale, or other salmon of poorer quality, may be designated by the inspectors as Grade B and this must be prominently declared on the PDP, cartons and cases.

The Regulations also specify that code markings are required for cartons and cases in which containers of fish are packed so that the name of the establishment and the day, month, and year of processing can be determined by an inspector. Cans packed in registered establishments must be embossed to similarly identify the establishment and processing date as well as the product therein. Hermetically sealed glass containers must bear the codes on the closure or label.

Meat Inspection Act

Administered and enforced by the Meat Inspection Division, Health of Animals Branch of Agriculture Canada, this Act and its Regulations control all foods containing meat, meat byproducts and meat derivatives which pass over provincial or international boundaries.

The Regulations specify that materials used in packaging a meat product that come in contact with the meat product must not impart any obnoxious quality to the meat and must be durable and effective when used. Meat must originate in a Health of Animals approved plant. Every dressed carcass or portion of a carcass passed by inspectors for food in these registered establishments and all packages or containers of meat or meat product leaving these registered establishments must bear or be labeled or tagged with the Inspection Legend which includes the registration number of the establishment. Food containing meat only in condimental proportions may be exempted. When intended for export, the required information may be in the language of the importing country. For imports, the words "product of" and the country of origin must be declared unless the country or origin is clearly displayed or indicated in an official stamp or brand of the national government. The abbreviation "Est." and the registration number required as part of the Meat Inspection legend, may be omitted from the Inspection Legend when it appears in other specified locations.

Standards are specified for many meat products and criteria of preparation are set out. Labeling requirements include that the name of ingredients may not be a collective name but must be a specific name, for instance, beef and beef tripe. Where more than one kind of meat except those known by a recognized name are used in a product, they must be described to indicate their exact nature and the different ingredients must be listed in the order of their proportion. When a flavor is indicated, it must be that of the meat used in the largest quantity. If commercial gelatin or any other permitted gelling agent is added to meat products, declaration of this addition is mandatory. Ham or bacon for export must be marked "Product of Canada." Perishable products in cans or similar receptacles must be marked "keep frozen" or "perishable, keep under refrigeration," as appropriate. Whale and horse meat and their products must be conspicuously labeled as such.

Canada Agricultural Products Standards Act

Administered by Agriculture Canada, this Act has Regulations governing a number of agricultural products. It establishes national standards and regulates international and interprovincial trade in these agricultural products.

Beef Grading Regulations and Veal Grading Regulations.—Once a carcass bears the Inspection Legend it is graded under these Regulations which specify grades and the grade and branding marks to be used. Carcasses that have not been graded may be exported or sent from one province to another for grading at destination.

Dressed and Eviscerated Poultry Regulations.— These stipulate that dressed or eviscerated poultry leaving the registered station, unless intended for further processing, must be squat packed, breast packed, or ice packed in approved and specified containers. Tolerances for bird uniformity are given. Individual birds may be wrapped in transparent material but the grade must be readily visible. Individual bird markings must contain:

(1) the name of the kind of bird, and in the case of turkeys, ducks, and geese, the word "Young" or "Mature" (as determined by criteria in the Regulations) preceding the kind;
(2) the name of the grade;
(3) the name and address of, or the assigned registration or certificate number of the registered station or certified producer grader, where the poultry was packed;
(4) in the case of poultry intended for export, the words "Product of Canada"; and
(5) the words "breast bone removed" if this is the case.

These declarations must be placed on metal tags, inserts, transparent bags, or other approved materials for packing individual birds. The metal tag must be securely attached to the birds at the "V" formed by the wishbone while the insert or bag marking must appear at the center of the breast. Where birds are packed in transparent bags that are closed by metal clips, the registration number may be engraved in an approved manner on the metal clip closure. Marking colors and sizes are specified, as are marking requirements for outer containers and imports.

Shell Egg Regulations.—Grades and standards for eggs and the packing of eggs are prescribed. Graders must place their mark of identification on the upper left hand corner of each grade tag, case, or box. Cases, boxes, or approved containers must contain fillers and flats or molded trays and must meet specified requirements of size and construction. Canada Grade A-1 eggs, must be packed in cartons of 6 or 12 (18, 24, or 30 egg cartons may be used with the Minister's approval) with a separate compartment for each egg and includes split-type cartons. Cartons may be packed in approved cases or boxes. Containers, except for those with separate compartments, must not contain more than one grade of egg. Eggs sold under a grade name other than Canada Grade A-1 are exempted from the Regulation that all eggs must be packed in cartons, cases, or boxes providing they are sold with the grade name prominently displayed on a card on top of or in front of the eggs.

Besides the usual declarations, cartons must be labeled with the name of the grade of the egg which includes the size designation. Cases and boxes may also require the weight designation and registered station number. Further declarations are required for imported eggs or eggs for export. The claims "new laid" or "fresh" may only be used for Grade A-1 and A eggs which have not been in cold storage.

Fresh Fruit and Vegetable Regulations.—The Fruit and Vegetable Division of Agriculture Canada administers the Fresh Fruit and Vegetable Regulations, Processed Fruit and Vegetable Regulations, Honey Regulations, and Maple Products Regulations. Products must be prepared in accordance with prescribed conditions and packed and marked in the prescribed manner. Most provinces have agreements with Agriculture Canada authorizing the federal inspector to act on their behalf for those products in intraprovincial trade.

The Fresh Fruit and Vegetable Regulations are concerned with fruits and vegetables which are grown in significant commercial volume in Canada such as apples, beets, lettuce, peaches, potatoes, beans, peas, corn, blueberries, grapes, etc. Fruits and vegetables not grown domestically (i.e., citrus fruits, bananas) do not come under these Regulations. Moreover, some aspects of packaging and labeling such as use of color or wax, or use of colored transparent containers come under the Consumer Packaging and Labeling and the Food and Drugs Acts.

Grades, standards, permitted condition defects, and tolerances are set out. With the exception of produce displayed in bulk in a retail store, grade names cannot be used in connection with produce unless it is packed:

(1) in standard packages as specified in the regulations or in approved experimental packages;

(2) in such a manner that the faced or shown surface represents the grade, size, maturity, color and variety of the produce contained in the package; and

(3) so that its quality or saleability is not adversely affected by the presence of insects, twigs, plant material, stones, or other foreign objects or material.

Apples, for instance, are graded as Canada Extra Fancy, Canada Fancy, Canada Commercial, Canada Commercial Cookers, Canada Hailed, Canada No. 1 (Peelers), and Canada No. 2 (Peelers). Besides other grade requirements, the Regulations set out tables of the color requirements for the grades Canada Extra Fancy, Canada Fancy, and Canada Commercial for red and red striped, red checked and blush, and green, yellow, and russet varieties. The apples must be sized with a minimum diameter of 2¼ in. and then packed according to the specifications set out in Table 10.3. The grade declaration on labels, tags, packages, or containers must be accompanied by the name of the apple variety, even for apples sold in bulk.

TABLE 10.3. PACKAGING REQUIREMENTS FOR APPLES

Pack	**Size or Weight**
Standard box	18 in. × 11 1/2 in. × 10 1/2 in.
Crate	17 in. × 13 in. × 11 in.
Carton	16 in. × 9 in. × 6 5/8 in.
Bag	3, 4, 5, 8 and 10 lbs
Half box	16 1/8 in. × 8 5/8 in. × 7 5/8 in.
Carton or crate	16 1/8 in. × 11 1/2 in. with a 5 3/4 in. end piece and a 4 3/4 in. side piece
Cell Pack	To be packed by count in individual cells of size:

	Count	Length	Width inches	Depth
(a) Elongated varieties	60	3 5/8	3 3/8	3 5/8
	72	3 7/16	3 3/16	3 7/16
	80	3 5/16	3 1/16	3 5/16
	96	3 1/16	2 7/8	3 1/8
	120	2 7/8	2 11/16	2 15/16
	140	2 11/16	2 9/16	2 3/4
	160	2 9/16	2 7/16	2 5/8
	175	2 15/32	2 11/32	2 1/2
	200	2 11/32	2 9/32	2 7/16
	216	2 7/32	2 7/32	2 3/8
(b) Round varieties	60	3 9/16	3 1/4	3 3/4
	72	3 7/16	3 1/16	3 9/16
	84	3 1/4	2 3/4	3 5/16
	96	3 3/16	2 11/16	3 1/4
	120	2 15/16	2 1/2	3
	140	2 3/4	2 5/16	2 13/16
	160	2 9/16	2 1/4	2 11/16
	180	2 1/2	2 3/16	2 5/8
	200	2 3/8	2 1/16	2 1/2
	216	2 5/16	2	2 7/16

There are many other specific Regulations governing the packaging and labeling of other products. If grade standards require that a package be marked to indicate the size of the produce therein, then the size declaration must be immediately adjacent to the grade name. For imports, a declaration of country of origin is required in addition to conformity to grade standards and all other packaging and labeling requirements. The kind of produce must be declared unless it is readily visible.

Processed Fruit and Vegetable Regulations.— These Regulations cover canned, frozen and dehydrated fruits and vegetables including those in sauces, pickles, relishes, jams, jellies, beans with pork, infant foods where fruit or vegetables are the major ingredient, and single strength juices and nectars. The few products not specifically regulated by this legislation fall under the Food and Drugs Act.

Grade standards and standards of identity are specified. Grade names established by these regulations may be applied only to products prepared in registered establishments, packed in prescribed standard containers to prescribed fills and marked as prescribed. Where the product fails to meet the lowest grade standard it may be marked "Sub Standard" providing the product is sound, wholesome and fit for food.

Accompanying the general label declarations must be, in both official languages:

(1) the grade name;

(2) a code mark clearly indicating the establishment where the product was packed and the date of packing if the product is a canned fruit or vegetable for which grades are established; and

(3) such other declarations are required by the Regulations.

The latter requirement covers such things as name of variety, style of pack, the word "Brand" below the trade name if it is a geographical location or some other descriptive name, the name of country of origin for imports, the size number or word designation for size graded products, sugar syrup declarations, etc.

As a rule of thumb, the net quantity declaration for canned products including jams and preserves is by volume measure; for frozen, dehydrated and glacéed products, it is by weight. Numbers for quantity declarations and letters for grade declarations must be of the same size; acceptable print sizes are specified.

All packages in which food containers are packed must be marked on one panel with the product grade, the number of containers in the package and the registration number of the establishment in which the product was packed.

Honey Regulations and Maple Products Regulations.— The Honey Regulations apply to all honey extracted from the comb and specify that honey must be packed, classified as to color, and graded in a registered establishment and marked according to the Regulations.

Consumer container sizes are specified as: up to 5 oz; 8 and 12 oz net; 1, 1½, 2, 3, 4, and 8 lb net. There are regulations for bulk container sizes. Each container must be securely closed by means of a screw cap, friction lid, bung, or other approved device. In addition to the standard declarations, the main panel of the label must declare in English and French:

(1) the word "honey" preceded by the type (i.e., "alfalfa," "honeydew") if known;

(2) the grade of honey, immediately followed by its color classification;
(3) the words "liquid," "pasteurized," "creamed," "pressed," etc., as appropriate; and
(4) the registration number of the packer included in the name and address declaration.

Print heights are specified depending on PDP size and must be the same for the grade, color and net quantity declarations.

"Product of Canada" or "Canadian" may be declared if this is the case. Imports must declare the name of the country of origin, preceded by "Product of." Bulk containers or box labels must also state the lot number.

Formerly under a separate Act, Regulations concerning maple products are now under the Canada Agricultural Products Standards Act. The Regulations are similar to those for honey. Labels must declare the license number, if any, of the manufacturer, packer, or sugar camp. Grade, color, and net weight markings must be of the same, specified print sizes.

Canada Dairy Products Act

New Dairy and Dairy Product Regulations are currently being promulgated but information sheets from the Dairy Products Division, Production and Marketing Branch of Agriculture Canada provide guidelines to the proposed packaging and labeling requirements. It should be noted that whole milk is not covered under this Act but is within the jurisdiction of the Food and Drugs Act and provincial legislation.

The labeling requirements of the Consumer Packaging and Labeling Act and the Food and Drugs Act apply and, in addition, the proposed Regulations will require bold face letters or numerals of the same specified heights on the PDP for:

(1) the net contents declaration;
(2) the numerals 1, 2 or 3 in the Maple Leaf grade mark;
(3) the numerals in the required percentage declaration of moisture and fat for butter spread, natural cheese, package cheese and any other dairy product not specifically provided for in the Regulations;
(4) all numerals and letters in the required percentage declaration of butter fat, vegetable fat, marine oil and other fat contained in butter-margarine blends; and
(5) the statement, when required, that a flavoring ingredient is either imitation, artificial or simulated.

For products packed in bulk, containers must also declare the churn, vat, batch or lot number, date of manufacture and registration number of the establishment.

There are some exceptions for the required dual language declaration of net contents and it is expected that dairy products packaged and labeled in the approved metric sizes will be exempted from the Canadian unit declaration.

The Regulations will contain new Maple Leaf grades and grade marks for many dairy products. These are indicated in Fig. 10.10 and must appear on the PDP. The Maple Leaf must conform with the official design and may be of solid color providing the printing is of a contrasting color.

For the ingredient listing, when a prepackaged dairy product is processed or made from basic materials (i.e., skim milk powder, natural cheese), the list of ingredients is headed by the words "made from"; when manufactured or consisting of many ingredients, the heading "Ingredients" is to be used. In general, class terms cannot be used but the class term "milk solids" is permitted for some ingredients.

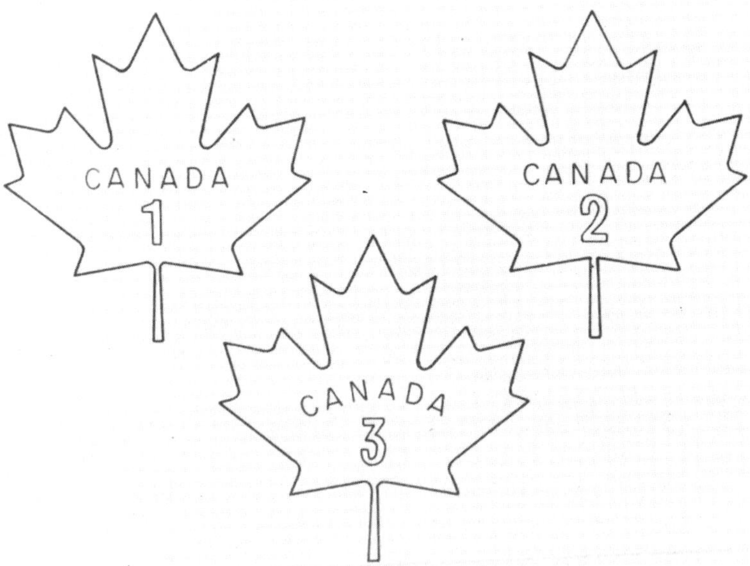

From illustrations in Agriculture Canada Dairy Division Information Sheet No. 8, January, 1976

FIG. 10.10. PROPOSED MAPLE LEAF GRADE MARKS FOR DAIRY PRODUCTS
The name "Canada" may appear in either a straight or curved line as indicated.

In line with proposals to make positive identification of the actual manufacturer necessary on all prepackaged foods, registration number requirements are being introduced in the new Dairy and Dairy Products Regulations. For hermetically sealed metal containers of dairy products, the registration number or an approved mark or code of the establishment in which they were prepared must be permanently embossed, printed or stamped on one end of the containers. Other containers must similarly be marked. Note that this requirement only applies when the name and address of the responsible party is not that of the actual establishment which prepared and packaged the product.

Specific standards, container size, and labeling requirements will be presented for a number of dairy products. The proposed Regulations will specify standards for the diffused transmittance of wrapping or packaging materials so as to end the oxidized flavor which is the most common flavor defect in print butter. For natural or process cheese, a grouped declaration of the maximum moisture and minimum fat content by weight will be required. Milk fat may be abbreviated "M.F." in the English declaration, "M.G." in the French. For skim milk powder, declaration of the name of the process (i.e., Instant) will be required. Special storage instructions such as "Keep Refrigerated" will be recommended for certain products even if the durable life date declaration is not required.

SPECIAL PRODUCTS

A number of products which have unique packaging and labeling requirement or which are of particular interest will be discussed in this section. Prepackaged products which are subject to regulations respecting packaging, labeling and marking under the Feeds Act, Fertilizers Act, Pest Control Products Act, and Seeds Act are exempt from many of the provisions of the Consumer Packaging and Labeling Act and thus are individual in their labeling requirements. The above four Acts are administered by the Plant Products Division, Production and Marketing Branch of Agriculture Canada. Special labeling requirements for cigarettes and the status of fluorocarbons in aerosols are of considerable current interest.

Cigarette Labeling

As in the United States, cigarette packages in Canada bear a warning on the label. It is noteworthy, however, that while the U.S. warning was made compulsory by their Federal Cigarette Labeling and Advertising Act of 1965, in Canada there is no such legislation. The declaration, in English and French: "Warning: Health and Welfare Canada advises that danger to health increases with amount smoked—avoid inhaling" and a declaration of tar and nicotine content on all cigarette packages is strictly voluntary on the part of manufacturers. The only rules governing cigarette label warnings are embodied in the Cigarette Advertising Code of the Canadian Tobacco Manufacturing Council, who do their own policing.

Environmental Contaminants Act

This Act was passed in 1976 by the Canadian government to control the use of certain chemical substances that are a threat to human health and/or the environment. The Act is the joint responsibility of Environment Canada and Health and Welfare Canada. The legislation is in place but regulations governing specific substances have not yet been promulgated. However, partially as a result of a report tabled in the House of Commons by the interdepartmental Atmospheric Environment Service which concluded that atmospheric release of fluorocarbons does cause a significant breakdown in the ozone layer, thus increasing risks of skin cancer, damage to plant and animal life, and climatic changes, regulations will soon be enacted to eliminate the nonessential uses of fluorocarbons F-11 and F-12 in aerosol products. Canadian aerosal manufacturers had voluntarily agreed to reduce by one-half the use of fluorocarbons by the end of 1977.

Feeds Act

This Act states that no feed shall be sold or imported into Canada unless the feed has been registered as prescribed, conforms to prescribed standards, and is packaged and labeled as prescribed. Any substance for consumption by livestock, for providing the nutritional requirements of livestock, or for preventing or correcting nutritional disorders of livestock is covered by the Feeds Act. Feeds for export, feeds not offered for sale, raw meat feed for mink and feeds sold by individual growers are exempted if the feed is free from prescribed deleterious substances such as DDT. Every feed must have attached to it or a package containing it a label or the shipping bill, in the case of feeds sold in bulk or for manufacturing, declaring, in both official languages, the following information:

(1) in the case of a feed not required to be registered, the name and address of the person who packaged or caused the feed to be packaged (when the feed was manufactured and labeled outside

Canada but the label declares the identity and principal place of business of a Canadian for whom it was manufactured, this name and address must be preceded by "imported by (for)" unless the geographic origin of the product is stated on the label);

(2) in the case of a feed required to be registered, the name, address, and registration number of the registrant;

(3) the brand of the feed, if any;

(4) the name of the feed, as approved or set out in the Regulations;

(5) if the feed is in the package in which it is to be sold, the net amount

  (a) in numbers of units in the package for capsules, tablets, or other individual doses or feeding units,

  (b) expressed as the net mass in metric units in accordance with the Weights and Measures Act for any other package of feed;

(6) in the case of a feed not required to be registered, the name of each ingredient;

(7) in the case of a feed required to be registered, the name of each ingredient or the statement "the ingredients in this feed are those named in the certificate of registration";

(8) an accurate statement of the guaranteed analysis (particulars required and tolerances are specified);

(9) if the feed is a mixed feed, directions for use;

(10) if the feed is in a form other than mash, the particular form of the feed;

(11) in the case of a medicated feed,

  (a) the name and actual amount of the medicating ingredient,

  (b) an acceptable precaution statement, and

  (c) a claim relating to the purpose of the medicating ingredient;

(12) in the case of a vitamin micro-premix, the lot number;

(13) if the feed is a tonic or conditioner,

  (a) the name of the disease or debility which the feed is intended to correct,

  (b) directions for use specifying the amount to be fed and that the feed is to be used only while the disease or debility exists, and

  (c) the word "tonic" or "conditioner" as part of the name of the feed;

(14) if the feed is a micro-premix or a macro-premix, the word "premix" must form a part of the name and if the feed is a supplement, the word "supplement" must form part of the name; and

(15) for mixed feeds, as part of the name or indirect association with the name,

(a) the kind of livestock or purpose for which it is intended (except that just "growing," "laying," "breeder," or a similar term is sufficient for feed intended for chickens), and
(b) a general nutritive classification of the feed (mineral feeds exempted).

Where there are inner and outer packages, both must be completely labeled. Customer-formula feeds have different label declaration requirements. There are a few other specific labeling requirements or restrictions. For instance, feeds of the same manufacturer which are identical but guarantee to contain different levels of protein, must have different brandnames or the percent protein as part of the name. With some exceptions such as retail sales from bulk and micro- and macro-premixes, packages of burlap, jute, cotton, paper, or similar materials must contain 5, 10, 20, 25, 40, and 50 kg of feed. It is of interest to note that a 1973 amendment to the Feeds Regulations prohibits the use of diethylstilbestrol in feeds.

Fertilizers Act

For the regulation and control of agricultural fertilizers, this Act specifies that no person shall sell or import into Canada any fertilizer or supplement unless it has been registered as prescribed, conforms to prescribed standards, and is packaged and labeled as prescribed. Packages must be labeled and bulk fertilizers must have statements or shipping bills which declare legibly and conspicuously, in either or both official languages:

(1) name and address of the manufacturer, registrant, or person for whom it was packaged;
(2) brand, if any, and name of the fertilizer;
(3) the net weight of the contents, in metric units;
(4) the registration number, where applicable;
(5) the guaranteed analysis prescribed in the Regulations;
(6) in the case of a fertilizer pesticide or a specialty fertilizer, directions for use;
(7) for fertilizer-pesticides, the name of the fertilizer shall include as a suffix the amount of the active ingredient of the pesticide in percent followed by the name of the pesticide;
(8) for fertilizer-pesticides, the prescribed cautionary statement;
(9) where the fertilizer is other than a specialty fertilizer and has intentionally incorporated in it or is represented to contain boron, copper, manganese, molybdenum, or zinc or, in the opinion of the Minister, has a natural high content of one or more

of these lesser plant nutrients, the following cautionary statement:
"CAUTION: This fertilizer contains ______ and should be used only as recommended. It may prove harmful if misused."; and

(10) for mixed fertilizers that contain any major plant nutrients, the grade designation as a hyphenated numerical series in whole numbers which represent minimum guarantees for total nitrogen, available phosphoric acid, and soluble potash in that order.

Other regulations specify labeling requirements for such things as customer-formula fertilizers and supplements. The word "manure" may not be used unless the product is guaranteed to contain at least 60% organic matter. Some fertilizers (i.e., those sold for correction of soil acidity only, natural manures) are exempt from the Act and its Regulations.

Pest Control Products Act

This Act regulates products used for the control of pests and the organic functions of plants and animals. Again, products must be registered as prescribed, conform to prescribed standards, and be packaged and labeled as prescribed. The Act demands that the registrant of a new pesticide supply the results of scientific investigations on the safety of the control product. All known scientific data are evaluated before registration is granted. Some products coming under the Food and Drugs Act and some devices are exempted. Furthermore, some control products are exempted from registration.

Where the product's primary purpose is pest control, the PDP shall show in English or French or both:

(1) the product name, which must be descriptive of form and purpose and include the common name of the active ingredient;
(2) the product class designation, "RESTRICTED," "DOMESTIC," or "COMMERCIAL";
(3) information respecting the degree and nature of hazard as in the Hazardous Products Act;
(4) the statement "READ THE LABEL BEFORE USING";
(5) a guarantee statement, expressed as "GUARANTEE," followed by specifications regarding the active ingredient;
(6) the registration number of the control product in the form "REGISTRATION NO. ______ PEST CONTROL PRODUCTS ACT";

(7) the net contents, in metric units only (effective January 1, 1980);

(8) the name and postal address of the registrant and the name and postal address of the resident agent, if any, of the registrant;

and the mandatory secondary display panel shall show:

(9) directions for use (dose rate, timing and use limitations);

(10) information identifying any significant hazard related to handling of the product;

(11) information identifying any significant hazard to public health, plants, animals, or the environment related to its use;

(12) instructions in first aid, under the heading "FIRST AID INSTRUCTIONS";

(13) toxicological information essential to treatment under the heading "TOXICOLOGICAL INFORMATION"; and

(14) "NOTICE TO USER — This control product is to be used only in accordance with the directions on this label. It is an offence under the Pest Control Products Act to use a control product under unsafe conditions."

Some devices (i.e., garment bags, electronic apparatuses) and products whose primary purpose is not pest control but which have such properties are exempted from many of these declarations. Products that are "RESTRICTED" must have, at the top of the secondary display panel, the "NOTICE TO USER, etc." declaration, followed by directions for use under the heading "RESTRICTED USES." This latter information must be set apart from all other label information by a circumscribed line. Information required on the label may be on a brochure or leaflet accompanying the product providing the display panel prominently states "READ ATTACHED BROCHURE (LEAFLET) BEFORE USING."

Declarations required for imports and bulk control products are also specified. The Regulations further stipulate that control products must be durably packaged to permit withdrawal of any or all of the contents in a manner that is safe to the user, to close adequately and to minimize degradation or changes of the contents. When the package is essential to the safe and effective use of the control product, it must be designed and constructed to specifications acceptable to the Minister on its registration.

Seeds Act

Grades and minimum standards of purity, germination, quality, and disease for seeds and their packing, marking, and package labeling are prescribed. Aside from the proviso that label declarations may be in

English or French, most of the labeling requirements are standard. Name and address of the seller, name of kind, variety and grade of the seed and name of country of origin if imported, net weight (for packages over 2 oz) and weed seed content declarations are required. For many seeds, besides the variety name, further statements clarifying the seed type are necessary (i.e., single-cut red clover). Packages containing more than 5 lb of seed must be marked with the lot designation. For many seeds and package sizes, use of variety names is restricted to seeds meeting certain grade standards. Moreover, seeds marked with a grade name must not have any statements or guarantees as to purity or percent germination unless detailed information on the name and number of noxious weeds and the correct percentage of germination is shown. Seeds of substandard grades must have declarations of percent germination and date when the germination test was completed. Seeds treated with pest control products must be stained a conspicuous color and have appropriate warning declarations.

There are exceptions and further specific regulations as to labeling for certain seeds, imports, and bulk seeds. Seed potatoes, with the exception of those sold in bulk, must be packed in containers that are new and closed after packing and have no marking that would lead to confusion with table potatoes.

PROVINCIAL LEGISLATION

Most provincial legislation which touches on packaging and labeling requirements reinforces federal regulations or is overridden by the requirements of the Consumer Packaging and Labeling Act. In some cases, however, provincial legislation has requirements not covered by federal regulations and in these areas, the provincial requirements have force. For instance, recycling of soft-drink and beer containers is controlled at the provincial level while grades, standards, and packaging requirements for fresh fruit and vegetables not covered in the Canada Agricultural Products Standards Act are often in provincial legislation. Furthermore, this federal Act pertains only to products for import, export, or interprovincial trade. In some cases, therefore, where provincial regulations may differ, adherence to provincial standards, grades, grade declaration, and packaging requirements is sufficient for products manufactured or produced and sold within that province.

As review of all provincial legislation which pertains to packaging and labeling is beyond the scope of this chapter, only a few examples of provincial regulations will be discussed.

Although increasing concern for the environment has been translated much more slowly into relevant recycling legislation in Canada than in the U.S., several provinces have recently passed legislation designed to divert glass from the solid waste stream. A 1976 amendment under the Ontario Environmental Protection Act requires every refillable container intended to contain carbonated soft-drinks for sale in Ontario must be clearly marked "MONEY-BACK BOTTLE—BOUTEILLE CONSIGNEE." The regulations restrict the use of non-refillable containers. Moreover, when the price of a carbonated soft-drink in a non-refillable cylindrical can or glass bottle is shown or announced in advertising or labeling, the price for the carbonated soft-drink in a refillable container of the same capacity must also be shown or announced.

In Quebec, the Official Languages Act of 1974 specifies all products must be labeled in the French language with no language taking precedence (i.e., in type size) over the French language in labeling. The Food Regulation under the Quebec Agricultural Products and Food Act specifies that the indication of origin of a Quebec product or one manufactured in Quebec must contain the word "Quebec." The indication of origin of all other products must be by the name of the country of origin or, in the case of a Canadian product, by the word "Canada," or the name of the province of origin or a similar expression. This indication of origin must accompany the price marking on a product, package, container, or placard. Similar declarations are required for fresh fruit and vegetables sold in bulk and all advertisements, announcements, or other publicity.

Several of the provinces have legislation pertaining to margarine. The Margarine Act of Saskatchewan defines standards for margarine and specifies that it must be labeled as such. The Margarine Act of Manitoba also specifies that no advertising respecting margarine and no label shall imply that margarine has a relation to any dairy product other than skim milk or depict a dairy scene.

TABLE 10.4. GENERAL CANADIAN LABELING REQUIREMENTS

To be declared in both English and French:

(1) Net quantity of product, in Canadian and metric units

(2) Identity of the product

(3) Identity and principal place of business of the person by or for whom the product was manufactured[1]

(4) The country of origin, if an import, or use of the phrase "imported by" when a Canadian distributor is identified

(5) For foods:
- (a) List of ingredients
- (b) Durable life date when less than 90 days
- (c) Grades and quality standards, where applicable

For drugs:
- (a) Quantitative list of medicinal ingredients
- (b) Lot number
- (c) Directions for use

For hazardous products:
- (a) Degree and nature of hazard symbols and signal words
- (b) Primary and secondary hazards
- (c) First aid instructions

For textiles[2]:
- (a) Textile fibre content and % of each fibre

[1] May be in either English or French
[2] Exempted from net quantity declaration

FURTHER INFORMATION

Table 10.4 provides a general summary of Canadian labeling requirements. For complete information on packaging and labeling in Canada, copies of the relevant Acts and their Regulations may be obtained from the:

Publication Center
Mail Order Services
Supply and Services Canada
270 Albert Street
Ottawa, Ontario
Canada, K1A 0S9

Further information in particular product areas may be obtained from the following:

Foods and Other Products, in General:

Director
Consumer Standards Directorate
Consumer and Corporate Affairs
Place du Portage
Ottawa/Hull
Canada K1A 0C9

Drugs, Cosmetics, and Medical Devices:

Director
Health Protection Branch
Health and Welfare Canada
Tunney's Pasture
Ottawa, Ontario
Canada K1A 0L2

Fish and Marine Products:

Director
Inspection Branch
Fisheries and Marine Service
Environment Canada
580 Booth Street
Ottawa, Ontario
Canada K1A 0H3

Agricultural Products:

Chief, Processed Products
Fruit and Vegetable Division
Production and Marketing Branch

Chief, Fresh Products
Fruit and Vegetable Division
Production and Marketing Branch

Director
Plant Products Division
Production and Marketing Branch

Director
Dairy Division
Production and Marketing Branch

Director
Meat Inspection Division
Health of Animals Branch of
Agriculture Canada
Sir John Carling Building
Ottawa, Ontario
Canada K1A 0C5

Environmental Contaminants:
Director
Environmental Protection Service
Environment Canada
580 Booth Street
Ottawa, Ontario
Canada K1A 0H3

In some cases, packaging and labeling approval is handled through regional offices.

ACKNOWLEDGEMENTS

The authors wish to express their appreciation to the members of various government departments for their cooperation and assistance. Special thanks are extended to Mr. Leo Leavy and Ms. Linda Arnold of the Department of Consumer and Corporate Affairs.

Index